Gabler Theses

AF148619

In der Schriftenreihe „Gabler Theses" erscheinen ausgewählte, englischsprachige Doktorarbeiten, die an renommierten Hochschulen in Deutschland, Österreich und der Schweiz entstanden sind. Die Arbeiten behandeln aktuelle Themen der Wirtschaftswissenschaften und vermitteln innovative Beiträge für Wissenschaft und Praxis. Informationen zum Einreichungsvorgang und eine Übersicht unserer Publikationsangebote finden Sie hier.

Felix Garayo Maiztegui

Design and Evaluation of an E-Learning Artefact for the Implementation of SAP S/4 Hana®

 Springer Gabler

Felix Garayo Maiztegui
Hard, Austria

This Ph.D. Thesis was accepted in 2022 by the Department of Strategic Management, Marketing and Tourism at the University of Innsbruck. Date of the Defense: 6th December 2022.

ISSN 2731-3220   ISSN 2731-3239 (electronic)
Gabler Theses
ISBN 978-3-658-40730-8   ISBN 978-3-658-40731-5 (eBook)
https://doi.org/10.1007/978-3-658-40731-5

© The Editor(s) (if applicable) and The Author(s), under exclusive license to Springer Fachmedien Wiesbaden GmbH, part of Springer Nature 2023
This work is subject to copyright. All rights are solely and exclusively licensed by the Publisher, whether the whole or part of the material is concerned, specifically the rights of translation, reprinting, reuse of illustrations, recitation, broadcasting, reproduction on microfilms or in any other physical way, and transmission or information storage and retrieval, electronic adaptation, computer software, or by similar or dissimilar methodology now known or hereafter developed.
The use of general descriptive names, registered names, trademarks, service marks, etc. in this publication does not imply, even in the absence of a specific statement, that such names are exempt from the relevant protective laws and regulations and therefore free for general use.
The publisher, the authors, and the editors are safe to assume that the advice and information in this book are believed to be true and accurate at the date of publication. Neither the publisher nor the authors or the editors give a warranty, expressed or implied, with respect to the material contained herein or for any errors or omissions that may have been made. The publisher remains neutral with regard to jurisdictional claims in published maps and institutional affiliations.

This Springer Gabler imprint is published by the registered company Springer Fachmedien Wiesbaden GmbH, part of Springer Nature.
The registered company address is: Abraham-Lincoln-Str. 46, 65189 Wiesbaden, Germany

To the living spirit

Acknowledgement

This research project would have been not possible without the orientation given by my two supervisors. First of all, I would like to thank my supervisors Prof. Priv. Doz. Dr. Frederic Fredersdorf and A. Univ.-Prof. Mag. Dr. Kurt Promberger, who orientated and challenged me to research, analyse, reflect, and keep involved in a continuous improvement cycle. Prof. Priv. Doz. Dr. Frederic and A. Univ.-Prof. Mag. Dr. Kurt Promberger coached me in different areas (e.g., adult training didactic, scientific theories and models, research methodologies and methods). In their tight agendas they always found a time slot for questions, feedback, and reflection rounds. They encouraged and motivated me. I also would like to thank the University of Innsbruck for the PhD seminars, upon we learnt ways of writing, several research methodologies, and good scientific practices. Additionally, I want to thank my wife Ingeborg Künz, who always had an open ear for my reflections, as well as understanding for our reduced free time together. Further, I would like to thank several members of the company Leica Geosystems, where the experiment was carried out. While Peter Ammann and Jürgen Sinz from the company Leica Geosystems AG allowed me to carry out the experiment for the validation of the hypotheses, the SAP trainers Ángels Xicoy, Evgenya Kaukhcheshvili, Mayte Castilla, and Kishore Kanthimahanti carried out trainings for the control group. Also, Hannes Töfferl integrated the SAP S/4 Hana sandbox system technically in the organisation and Adam Mark Miller proofread the dissertation. Finally, I would like to express my thanks to all employees of the company Leica Geosystems worldwide who answered the questionnaires and participated in the experiment.

Abstract

This research project aims to create a new e-learning artefact for SAP S/4 HANA training purposes. Based on Design Science Research, a new prototype of an e-learning artefact has been developed. The prototype is grounded on didactic and information systems theories, as well as models. The aim of this new prototype is to transfer knowledge of business processes with the new Enterprise Resource Planning (ERP) SAP S/4 HANA effectively and efficiency. The study was validated through an experiment with a treatment and a control group. While the treatment group learnt with the new e-learning artefact, the control group kept with the existing conventional training. Results show that the treatment group had significantly less dropouts than the control group, and that the treatment group could also finalise the SAP S/4 HANA exercises in less time. This study shows also how companies can benefit with this study using the same approach through a set of guidelines.

Abstract in German language

Dieses Forschungsprojekt zielt darauf ab, ein neues E-Learning-Artefakt für SAP S/4 HANA-Schulungszwecke zu erstellen. Basierend auf Design Science Research ein neuer Prototyp eines E-Learning-Artefakts ist entwickelt worden. Der Prototyp basiert auf didaktischen und informationstechnischen Theorien sowie Modellen. Ziel dieses neuen Prototyps ist es, das Wissen über Geschäftsprozesse mit dem neuen Enterprise Resource Planning (ERP) SAP S/4 HANA effektiv und effizient zu übertragen. Die Studie wurde durch ein Experiment mit einer Experimental- und einer Kontrollgruppe validiert. Während die Experimentalgruppe mit dem neuen E-Learning-Artefakt übte, lernte die Kontrollgruppe mit bestehend konventionellen Schulungen. Die Ergebnisse zeigen, dass die Experimentalgruppe weniger Kursabbrecher als die Kontrollgruppe hatte, und dass die Experimentalgruppe die SAP S/4 HANA-Übungen auch in kürzerer Zeit abschließen konnte. Diese Studie zeigt auch, wie Unternehmen von dieser Studie profitieren können, indem sie den gleichen Ansatz durch ein Vorgehensmodel verwenden können.

Contents

Abbreviations

ARIS	Architecture Integrated Information Systems
BPMN	Business Process Model and Notation
BOM	Bill of Materials
CAD	Computer Aided Design
DSR	Design Science Research
ECC	ERP Central Component
EPC	Event-driven process Chain
GPS	Global Positioning Systems
GTCs	General Terms and Conditions
GUI	Graphical User Interface
HANA	High Performance Analytical Appliance
IPMA	International Project Management association
IoT	Internet of Things
ISSM	Information System Success Model
IT	Information Technology
KPIs	Key Performance Indicators
LMS	Learning Management System
LoBs	Line of Businesses
OiH	Orders in Hand
OTS	Orders to Cash process
PC	Personal Computer
PHP	Hypertext Preprocessor, originally known as Personal Page Home Tools
PMA	Project Management Austria
SaaS	Software as a Service
SAP®	System, Application, Processes

SCORM	Sharable Content Object Reference Model
SOP	Sales Operations Planning
UML	Unified Modelling Language
UX	User Experience
VPN	Virtual Private Network

List of Figures

List of Tables

Introduction and Motivation for the Topic

1

1.1 Background and Initial Situation

Business processes are business managerial activities that regularly repeat (Gadatsch, 2015, p. 3). Business processes have an input as well as an output, and they should add value to companies' business itself. Business processes are basically, in a detailed form, workflows. Workflows contain activities with information in terms of time, professional activities and resources (Gadatsch, 2015, p. 5). Effective and efficient business processes contribute to the economic success of a company, i.e., with a better quality of the products, with optimised costs in logistics or with an excellent customer service to achieve a better customer satisfaction. The modelling of business processes supports the graphical representation of workflows i.e., in the fields of sales, after sales as well as services (Gadatsch, 2015, p. 3). In this sense the representation through flow charts helps employees to understand and learn workflows of business processes. These workflows in a software system environment can be manual, semi-automatic or fully automatic. The more automatic the workflow is, the more resources will be needed in terms of Information Technology (IT) development and IT resources. There are several ERP (Enterprise Resource Planning) software systems to support the representation and data processing of business processes in companies (Körsgen, 2000, p. 2). One of the most known ERP systems in the market is System, Applications and Processes (SAP). ERP SAP® has a market share of 44% in EMEA (Europe, Middle East, and Afrika) and therefore is one of the leading ERP software providers worldwide (das Statistik-Portal, 2017, para. 2). Organisations should apply business processes and workflows actively. This requires, as stated by Scherrer and Schaffner (2003, p. 59), among other factors, also a proper training of employees. The objective of a training does not only mean the

© The Author(s), under exclusive license to Springer Fachmedien Wiesbaden GmbH, part of Springer Nature 2023

F. Garayo Maiztegui, *Design and Evaluation of an E-Learning Artefact for the Implementation of SAP S/4 Hana®*, Gabler Theses, https://doi.org/10.1007/978-3-658-40731-5_1

understanding of business processes and workflows, but also the effective and efficient utilisation of SAP ERP software systems. Especially in the area of software services, where the software development level is high, it is a challenge to retain and improve the knowledge level among employees. This is a complex and difficult task, mostly because the resources in companies to keep that knowledge level are not always available (Paa, 2014, p. 4).

1.2   Problem Formulation and Motivation

There are several concepts available in the market for modelling processes. According to Gadatsch (2015, p. 16) well known concepts are: Business Process Model and Notation (BPMN), Unified Modelling Language (UML), Activity Diagram, Event-driven Process Chain (EPC), Swimlane-Diagram and Value-added Chain Diagram. In the case of SAP ERP the most common modelling method used is the so-called Architecture Integrated Information Systems (ARIS). It includes notations of BPMN and EPC (Gadatsch, 2015, p. 42). These two types of notations enable the representation of different workflows from five perspectives like Organisation, Function, Data, Output and Control. ARIS includes an own modelling concept and tool set. SAP ERP offers with ARIS the possibility of automatising workflows for the technical implementation of several SAP modules. Such SAP modules are available, for example, in the areas of Logistics, Finance and Controlling (Rimmelspacher, 2014, p. 2). Other classical well-known logistic modules are Sales and Distribution (Scheibler, 2002, p. 33) or Customer Service, where the automatism of workflows in the context of sales and service processes at companies are also supported (Oswald, 2003, p. 25).

ERP SAP itself is a relational database and a comprehensive software system solution with more than 100.000 (one hundred thousand) and, theoretically speaking, unlimited number of tables. The Graphical User Interface (GUI) of SAP ERP is a representation with hundreds of fields accessing information to these tables. Knowledge transfer in terms of teaching dozens of standard and customisable workflows, as well as of SAP transactions to call the needed programs to perform those workflows, require a lot of resources for any organisation. This also includes the subsequent support for those employees carrying out tasks in the SAP ERP system. According to Paa (2014, p. 1) companies should spend in an SAP ERP roll-out around 15% of the total SAP ERP implementation budget to properly train employees. The effective training of employees plays an important role to achieve a successful and cost-efficient implementation of an ERP SAP system (Körsgen, 2001, p. 10). SAP ERP provides nowadays, and since many years, courses and trainings for the different modules mention above. These courses are also available on an internet-based online learning platform (SAP AG, 2017).

However, these courses do not satisfy many companies' needs when it comes to training SAP ERP. The reason for this is the continuous SAP ERP system customization and programming for new, as well as adapted workflows based on customer needs (Teufel et al., 2000). If no proper training of employees does happen, there will be also errors in the order entry with subsequent employees' frustration and customers' complaints. These customers' complaints might also have an impact on future revenues. The low operating costs strategies as well as the limited personnel resources are in several companies working with SAP ERP also a challenge, and as a result, many times there is very little or no budget left for SAP ERP trainings, especially when it comes to teaching improved workflows (Paa, 2014, p. 4).

SAP S/4 HANA stands for "SAP Business Suite version 4 High Performance Analytical Appliance" and represents, in comparison to the current existing SAP ERP, a new in-memory technology and an optimised program code to support companies towards the new digitalization that is currently happening (Preuss, 2017, p. V). SAP S/4 HANA versus SAP ERP enables companies to reach quick data in real-time, to react quickly to changes in business environments, to optimize the profit of companies, as well as to prepare them for a digital transformation in the future (Burgdorf, J., Destradi, M., Kiss M, Schuber, M., 2017, p. 19). SAP S/4 HANA also gives the possibility to run business reports that were until know unthinkable with SAP ERP (Brenckmann & Pöhling, 2013, p. 26). The graphical user interface (GUI) of SAP S/4 HANA in comparison to SAP ERP is also new, and to a certain extent, based on so-called SAP Fiori® apps. With these Fiori apps it is possible to achieve a better personalisation user experience (Burgdorf et al. 2017, p. 20). This represents not only a new GUI for the user, but also a new way to manage workflows, where companies will change the way that they work with their data. With SAP S/4 HANA several business processes and workflows can be processed through Fiori Apps. Fiori Apps can simplify the processing of specific business processes and workflows in terms that only needed process relevant fields and information are shown. The question how to transfer knowledge of business processes and workflows with SAP Fiori apps utilising a new e-learning prototype represents the major problem formulation and motivates the topic for this dissertation project.

1.3   Objective and Research Question

The objective of this dissertation thesis is to create, develop and scientifically validate the prototype of a new learning artefact for transferring knowledge of business processes and workflows with SAP S/4 HANA. An artefact is a form of a construct, a model, a method, or an instantiation (Hevner, March, Park, & Ram, 2004, p. 78). While constructs are the language to define problems and solutions as well as to communicate, models use constructs to represent problems and their potential solutions (Hevner et al., 2004, p. 78). Methods define the process, how the problems through potential solutions can be solved: i.e., best practices solutions. Finally, instantiations show how constructs, models or methods can be implemented (Hevner et al., 2004, p. 79). An artefact for this dissertation thesis is an instantiation. The e-learning artefact consider the business processes and workflows for software services and the output of this dissertation project should be a validated technological concept in terms of a prototype. This prototype could be used later for other business processes and workflows for SAP S/4 HANA. This dissertation is written in English language with the aim of making the content available to several companies in a maximum possible number of countries.

The research of the dissertation thesis concerns the conceptual and empirical validation of a new learning artefact for a know-how transfer of business processes and workflows using the SAP S/4 HANA as an integrated business process software. This research should provide a new approach in terms of teaching new and existing business processes for the SAP S/4 HANA solutions for software services. The following research questions should be answered:

- In what way is an e-learning artefact suitable to transfer knowledge of business processes using a new SAP S/4 HANA in the area of software services?
- In what way is an e-learning artefact suitable to transfer knowledge of improved business processes using a new SAP S/4 HANA in the area of software services?

1.4   Structure of the Dissertation

The structure of this dissertation begins with a chapter related to Design science. Design science aims to create knowledge about how to develop and design solutions (inner environment) to solve problems in society or organisations in terms of outer environments (Dresch, Pacheco, Valls, 2015, p. 56). Design science is also the scientific basis for the development of the e-learning artefact for this research project and provides the needed methodological and methodical structure. It starts with the environment analysis where the e-learning artefact should be implemented (chapter 3). The environment refers to the context of the company Leica Geosystems itself, the requirements for the training of SAP S/4 HANA processes, and the description SAP S/4 HANA processes. The descriptions of the SAP S/4 HANA processes are important, because although the prototype of the e-learning artefact will cover the sales order process for software services, it is also the intention to utilise this e-learning artefact for other SAP S/4 HANA processes (e.g., production, technical service, purchasing, procurement, warehousing, finance, controlling, etc.). The structure continues with the knowledge base (chapter 4) needed in terms of scientific knowledge and expertise for the development of the e-learning artefact. The scientific knowledge comprehends didactic theories and models, as well as the requirements for developing e-learning artefacts from different scientific models like ISSM (Information System Success Model). The environment as well as the knowledge base represent the core elements for the design and development of the e-learning artefact (chapter 5). Chapter 5 also shows the empirical research design for the operationalisation of DSR, and includes the hypothesis, variables, and indicators. Further, in chapter 6 there is an evaluation of the experiment carried out for the validation of the hypotheses. The evaluation looks to effectiveness and efficiency related results. Because the e-learning artefact could be utilised for other SAP S/4 HANA business processes, chapter 7 shows a guideline how a company should proceed at the time of implementing the e-learning artefact. This dissertation ends up with a conclusion related to the theoretical and empirical part, as well as the implications in the practical world. Finally, it also shows the limitations of the study.

Research Design

2

2.1   Research Design and Methodological Approach

An artefact or Information Technology (IT) artefact has been defined in literature as an Information System by many researchers (Offermann, Blom, Schoenherr, Bub, 2010, p. 77). The design and implementation of an artefact is an Information System initiative in terms of an intervention of a social-technical system (Hevner, & Chatterjee, 2010, p. 219). According to Dumay and Baard (2017, p. 267) Interventionist research (IVR) is a research approach where the researcher seeks to determine an experimental situation, acts in that situation within an organisation immersed with the object of study, and analyses findings in view of relevant literature (Jönsson, & Lukka, 2007, p. 374). In IVR the researcher is an active actor in the research project of a specific research field (Jönsson, & Lukka, 2007, p. 374). IVR is an "umbrella" for several synonym research designs approaches like action research, action science, design science, clinical research, and constructive research (Jönsson, & Lukka, 2007, p. 376). Design science as IVR has its origin in the book "Science of the artificial", also formerly known as Science of Design (Simon, 1996, p. 111). Design science aims to intervene in environments to make improvements, as well as to design solutions to problems and thus develop knowledge in the context of applied research (Saunders et al., 2009, p. 9). The mission of Design science as IVR is to develop knowledge through intervention for the design and realisation of IT artefacts (van Aken, 2004, p. 224). Design science is oriented towards the future and is per nature interventionist (McKay, Marshall, Heath, 2010, p. 113). In this sense the planned research aims through intervention also to develop knowledge in terms of how

© The Author(s), under exclusive license to Springer Fachmedien Wiesbaden GmbH, part of Springer Nature 2023
F. Garayo Maiztegui, *Design and Evaluation of an E-Learning Artefact for the Implementation of SAP S/4 Hana®*, Gabler Theses,
https://doi.org/10.1007/978-3-658-40731-5_2

an e-learning artefact can be built up for transferring knowledge of business processes in an effective way with SAP S/4 HANA as an ERP system. The researcher
is also engaged with the research project.

The research philosophy, in terms of methodological approaches of the IVR,
is important for making assumptions about the ways the world is seen as well as
how knowledge is developed (Saunders et al., 2009, p. 108). These assumptions
will underpin the research design in terms of the methodological fit elements
like type of data collected, illustrative methods of collecting data, constructs and
measures, goal of data analysis, data analysis methods as well as the contribution to the literature (Edmondson, & McManus, 2007, p. 1156). There are
few researchers who have done some attempts to position the Design science
in the critical realism (Carlsson, 2006, p. 1) and interpretivism methodological approaches (Niehaves, 2007, p. 1). However, none of these attempts clearly
specify the epistemological grounds for the useful nature of design knowledge
(Goldkuhl, 2011, p. 85). Simon (1996, p. 135) as the founder of Design science
as well as Hevner & Chatterjee (2010, p. 181) state that Design science as IVR is
rooted in the methodological approach of Pragmatism. For this research project
I also believe that Pragmatism is the most adequate methodological approach. In
the following the ontological and epistemological foundations of the IVR related
to the Pragmatism are described. While ontology concerns the issue, what exists
or also what is the fundamental nature of reality, epistemology is the area of soci-
ology that concerns the creation of knowledge related to the explanation of social
reality, whether an explanation is true or false, and what does good evidence looks
like (Neumann, 2014, p. 96).

According to Cresswell (2014, p. 294), Pragmatism as a philosophy or world-
view arises out of actions, situations, and consequences, emphasising the research
problem as well as the use of all available approaches to understand it. Pragma-
tists believe in an external world independent of the mind, as well as lodged
in the mind (Cresswell, 2014, p. 40). For Pragmatism reality may vary between
realism or nominalism ontologies (Burrell, & Morgan, 1979, p. 4). Realism postulates that the social world is made of structures that exists as empirical entities
and that the individuals live in a world that has its own reality independent of the
researcher (Burrell, & Morgan, 1979, p. 4). For the realistic ontology the world
exists "out there", is patterned, and has a natural order (Neumann, 2014, p. 98).
The nominalism ontology assumes that the social world is *external to individual
cognition made up of names, concepts and labels which are used to structure reality*" (Burrell, & Morgan, 1979, p. 4). Hevner and Gregor (2013, p. 337) postulate
that design science relies on a pluralistic form of realism with three domains:
world 1 (objective world of material things in terms of instantiations), world 2

(subjective world of mental states like e.g., design theories), and world 3 (objectively existing but abstract world of human-made entities in terms of constructs). Design science, especially in engineering (an IT artefact belongs to the software engineering disciplines), adopts a realistic ontological position of Pragmatism because it assumes that instantiations are part of an objective world as a material thing, and that organisations (e.g., companies) are structures that exist, where employees work. Also, these organisations have their own reality independent of the researcher.

Korte and Mercurio (2010, p. 71) affirm that in Pragmatism, ideas and knowledge are viewed as tools created by people through their experiences and that researchers can use experiences as important tools to get things done in the world. In Pragmatism, based on the research question, it is possible to work within positivist and/or interpretivist positions (Saunders et al. 2009, p. 598). According to Cresswell (2014, p. 40) pragmatism looks to what and how to research based on the intended consequences. Korte and Mercurio (2010, p. 71) state that to produce knowledge in Pragmatism the tenet of rigorous inquiry should be considered. Rigorous inquiry starts with an indeterminate situation that causes to the researchers to doubt something, from this doubt they formulate a question, further they create potential solutions, they test these solutions via experimentation (positivist approach) and gather empirical evidence (Korte, & Mercurio, 2010, p. 64).

2.2 Design Science

The concept of Design Science has its origin in the book "Science of the artificial" where originally was defined as Science of Design (Simon, 1996, p. 111). Design focuses on how things, through the development of artefacts, should be (Simon, 1996, p. 114). According to Simon (1996, p. 114) an artefact *"can be thought as a meeting point (in the sense of an interface), between an inner environment (the substance and organisation of the artefact itself) and the outer environment (the surrounding in which operates)"*. Design science aims to create knowledge about how to develop and design solutions (inner environment) to solve problems in society or organisations in terms of outer environments (Dresch, Pacheco, Valls, 2015, p. 56). Design means to invent and bring-into being a new artefact that does not exist (Vaishnavi, Kuechler, & Petter, 2018, p. 3), being this one innovative. This innovation may also mean to carry out research with the purpose of solving problems and thus try to fill knowledge gaps (Vaishnavi, Kuechler, & Petter, 2018, p. 3).

Design Science aims to create new knowledge through the design of innovative artefacts and to analyse them with reflection and abstraction (Vaishnavi,
Kuechler, & Petter, 2018, p. 1). Table 2.1 below shows an overview about the
differences of natural sciences, social sciences and design sciences in terms of
purpose, research goal and some scientific areas. The starting point of Design
Science usually starts with the need of designing an artefact (Dresch, Pacheco,
Valls, 2015, p. 61); e.g., a software for learning purposes. Its operationalisation
requires a research method to carry it out with rigor and to validate it scientifically (Dresch, Pacheco, Valls, 2015, p. 57). Design Science Research (DSR) is
the method for this research project.

Table 2.1 Natural, social and design science (Dresch et al.,2015, p. 14)

Characteristic	Natural science	Social science	Design science
Purpose	To understand complex phenomena. To discover how things are and to justify why they are this way.	To describe, understand, and reflect on human beings and their actions.	To design; to produce systems that do not exist; to modify existing situations to achieve better results. Focus is on solutions.
Research goal	To explore, to describe, explain, and predict.	To explore, describe, explain, and predict	To prescribe. Research is oriented towards solving problems.
Examples of areas that usually employ each of these scientific paradigms	Physics, chemistry, biology.	Anthropology, economics, politics, sociology, history	Medicine, engineering, management.

2.3   Design Science Research Method

DSR is a firmly established research method for the design of an artefact to solve
real-life problems (Fischer, 2011, p. 2). The idea of the Design Science Research
(DSR) in Information Systems is to aid to the productive utilisation of information
systems for individuals, groups, and organizations (Hevner et al., 2004, p. 76);

e.g., a new e-learning artefact for the implementation of a new ERP (Enterprise Resource Planning) system. Design science research should create a viable arte-fact in the form of a construct, a model, a method, or an instantiation (Hevner et al., 2004, p. 78). While constructs are the language to define problems and solu-tions as well as to communicate, models use constructs to represent problems and their potential solutions (Hevner et al., 2004, p. 78). Methods define the process, how the problems through potential solutions can be solved: i.e., best practices solutions (Hevner et al., 2004, p. 79). Finally, instantiations show how constructs, models or methods can be, functionally speaking, implemented (Hevner et al., 2004, p. 79). Design Science Research (DSR) is based on seven guidelines for solving a problem. These guidelines are design of an artefact, problem relevance, design evaluation, research contributions, research rigor, design as a research pro-cess and communication of the research (Hevner, March, Park, & Ram, 2004, p. 83). The main objective of the *problem relevance* is to acquire knowledge for the development of technology-based solutions for unsolved important business problems by creating innovative artefacts (Hevner et al., 2004, p. 84). Accord-ing to Herver and Gregor (2013, p. 351) any *design evaluation* of an artefact is submitted to a validity, utility, quality, and efficacy through consistent eval-uation methods that are determined by the business environment including its technical infrastructure. An additional guideline is the *research contribution* of the DSR, this means, what are new and interesting contributions of the research (Hevner et al., 2004, p. 87). The DSR research must contribute to three areas: the design of the artefact, the design construction knowledge (i.e., foundations) and the design evaluation knowledge in terms of methodologies (Hevner et al., 2004, p. 87). Also relevant is the *research rigor*, this means, the rigor utilizing scientific methods when creating and evaluating artefacts (Hevner et al., 2004, p. 87). DSR is like the learning; it's not a thing but a process. In this *process* there should be means that reach the desired ends and build the needed knowledge platforms (Hevner et al., 2004, p. 88). DSR should be *communicated* not only in scientific environments but also in the environments where they should be applied; (Hevner et al., 2004, p. 90) i.e., in companies or other types of organisations.

2.4   Three Cycle view of Design Science Research

Based on these seven guidelines mentioned, Hevner (2007, p. 87) published a three-cycle view of a DSR framework. The three-cycle view of the DSR (see Figure 2.1) takes into consideration three cycles: relevance, design and rigor

(Hevner, 2007, p. 88). The *relevance cycle* is the interface between the environment and the DSR activities. The motivation of the relevance cycle is to improve the environment in terms of introducing innovative artefacts (Hevner, 2007, p. 88). The environment takes into consideration the people, the organisational and the technical systems. Thus, DSR can support the development and the construction of artefacts as well as strengthen the existing knowledge base (Dresch et al., 2015, p. 68). The relevance cycle includes the requirements based on the problem found but also the needed acceptance criteria for the evaluation of the results through field testing (Hevner, 2007, p. 89). The results of the field testing will determine if additional iterations of the relevance cycle are needed (Hevner, 2007, p. 89).

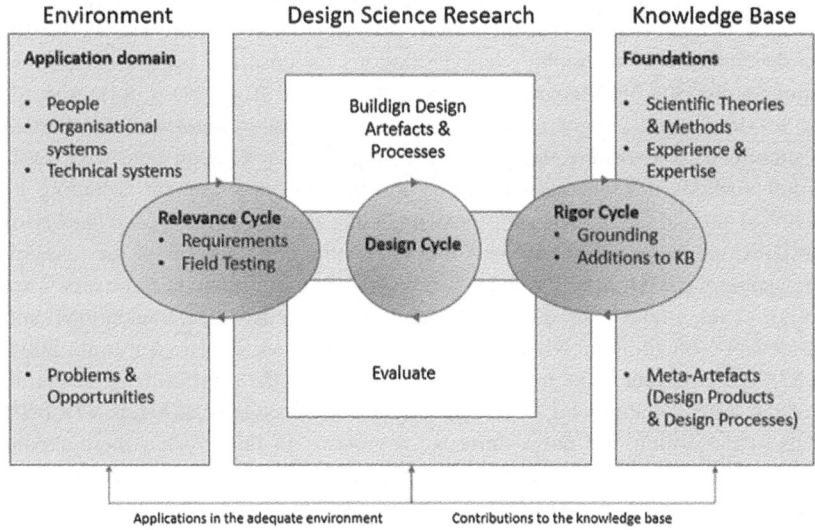

Figure 2.1 The Three Cycle View of Design Science Research. (Source: adapted from Hevner, 2007, p. 88)

The *rigor cycle* interacts with the Design Science activities and contributes with knowledge in terms of scientific theories and methods that provide foundations for rigorous DSR (Hevner, 2007, p. 89). These foundations are the scientific theories and methods as well as the experience and expertise. The idea behind is that researchers use existing scientific knowledge to guarantee that the designs produce research contributions also to this scientific knowledge (Hevner, 2007,

p. 90). This could also mean the creation of meta-artefacts, this means the development of existing methods and theories for the creation of design products and processes (Hevner, 2007, p. 90). Hevner and Gregor (2013, p. 344) stated that the contribution to the knowledge can be different depending on starting points related to problem and solution maturity levels (see Figure 2.2). An invention is a new solution for a new problem where none or very little understanding exists and where no effective artefacts have been created yet (Hevner, & Gregor, 2013, 346). While for an invention the problem is unknown for an improvement the problem known. An additional differentiation to the invention is that an improvement represents a contribution to the knowledge that lays in the development of a new solution for a clear suboptimal problem (Hevner, & Gregor, 2013, p. 346). Exaptation means the extension of known solutions to new problems, whereas the researcher needs to demonstrate, that the knowledge into the new field is interesting. (Hevner, & Gregor, 2013, 347). A routine design would not mean a major knowledge contribution, because it means to apply known solutions to known problems. *The design cycle* is the core of the DSR and interacts among the *"construction of the artefact, its evaluation, and subsequent feedback to refine the design further"* (Hevner, 2007, p. 90). In this cycle it is important not only to have a grounded reason for the construction of the artefact based on scientific defined requirements, but also to evaluate rigorously such an artefact (Hevner, 2007, p. 91). This dissertation is innovative in terms of an improvement (new solutions for known problems) and will consequently contribute to the literature. It is especially this innovation and the rigor what distinguish Information systems as design science research from the practice of building general IT artefacts (Livari, 2007, p. 41).

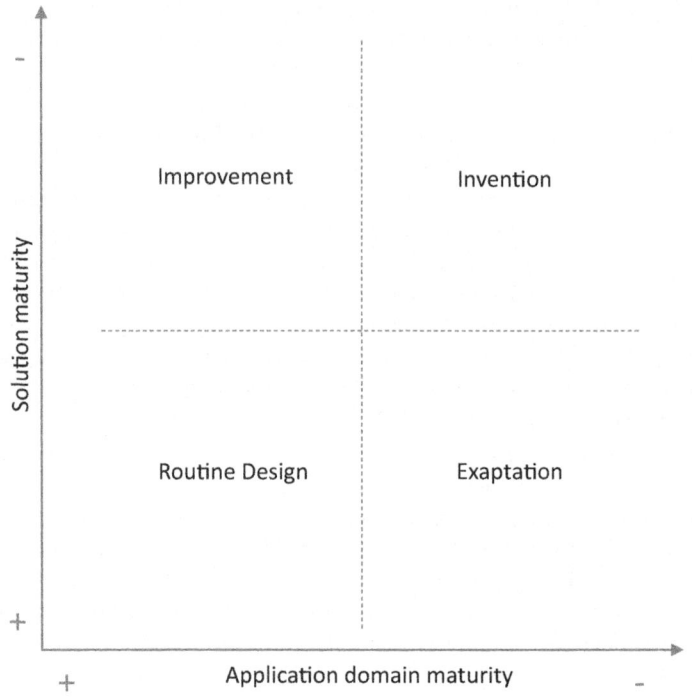

Figure 2.2 DSR knowledge contribution framework (Source: Adapted from Gregor, Hevner, 2013, p. 345)

2.5 Design Science Research in the Context of this Dissertation

According to the Three Cycle View of DSR, the environment for the application domain of this research project refers to the context of the company Leica Geosystems AG itself, the requirements for the training of SAP S/4 HANA processes, and the description SAP S/4 HANA processes as technical systems (see Figure 2.3). The knowledge base is related to the didactic theories and models, the requirements for developing e-learning artefacts from different scientific models like ISSM (Information System Success Model), and the knowledge related to the design of artefacts. The design cycle is based on both the environment as well as the knowledge base (see figure 2.3). The next chapter will start with the environment. Prior to the development of any e-learning artefact it is important to describe in which context the e-learning artefact is needed, and also the requirements of the company Leica Geosystems for such an e-learning artefact. These requirements contain different perspectives (e.g., financial, organisational requirements) and provide a framework for the later development of the artefact. The environment will also consider the SAP S/4 HANA Processes. The company Leica Geosystems will migrate from SAP ERP into SAP S/4 HANA in April 2023 through a "grey" approach, this means that SAP will migrate to the new SAP HANA® database. The users at the beginning will not realise any changes, however in the following months SAP Fiori apps will be developed not only for existing, but also new SAP S/4 HANA business processes. This is the reason why the environment as part of the DSR also considers SAP S/4 HANA processes and technical ERP systems.

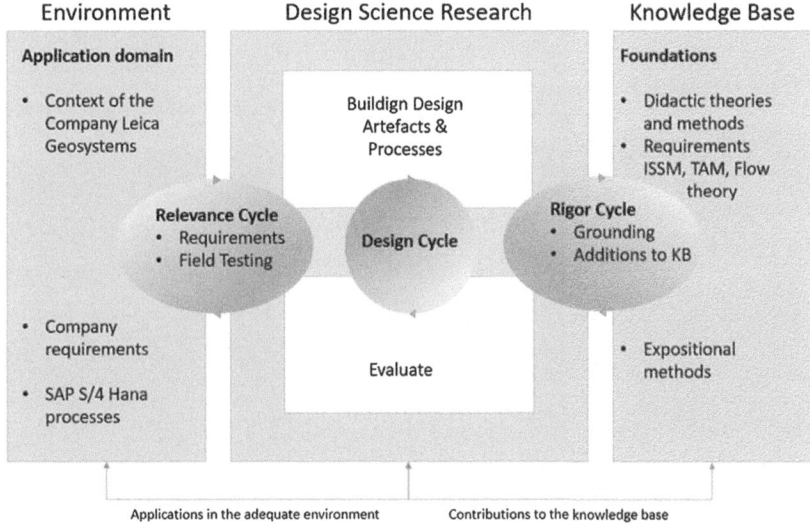

Figure 2.3 DSR knowledge contribution framework for this research project (Source: Adapted from Hevner, 2007, p. 88)

Environment

3

The environment of the DSR model for this research project considers the context of the company Leica Geosystems AG in terms of a general description, the requirements for teaching SAP S/4 HANA and the SAP S/4 HANA processes that will be implemented after the migration to SAP HANA in April 2023.

3.1 Context of the Company Leica Geosystems AG

The company Leica Geosystems AG is a worldwide company that provides equipment, software, services, and solutions mainly for surveying, construction, and mining industries.

The company belongs to the Hexagon group, a Swedish concern with stock exchange in London. Leica Geosystems AG has several own sales, support and technical services departments in Europe, Middle East, Asia, and North America. In other parts of the world Leica Geosystems sells and distributes its products and services through distribution partners. Most of these distribution partners are in South America, Middle and Far East. Leica Geosystems AG counts with 24 own sales organisation units worldwide that currently work on daily basis with SAP as their ERP system. Leica Geosystems AG has four big distribution centres in Germany, Switzerland, United States of America, and Singapore. The company counts with approximately 3000 employees worldwide. The SAP system runs under one single mandate. This is beneficial and crucial for consolidating the profit and loss contributions into one general ledger. However, it also represents a big challenge, because the customer requirements might vary from continent to continent, and from country to country. These variations cause permanent adaptations in the general business processes and subsequently in the SAP business

© The Author(s), under exclusive license to Springer Fachmedien Wiesbaden GmbH, part of Springer Nature 2023
F. Garayo Maiztegui, *Design and Evaluation of an E-Learning Artefact for the Implementation of SAP S/4 Hana*®, Gabler Theses, https://doi.org/10.1007/978-3-658-40731-5_3

processes too. The fact that Leica Geosystems AG sells hardware, software, services, and solutions worldwide means also that many sales, support and service models need to be supported with the same SAP ERP system under one mandate. For example, the company sells the same surveying software under the license model permanent, time limited and subscriptions. The SAP processes in the background are quite complex and the proper handling requires a significant support as well as continuous training of employees.

3.2   Requirements of the Company

The company Leica Geosystems AG plans to migrate to SAP HANA in April 2023. The migration is called a "bluefield" approach. This means that the configuration of SAP ERP will be migrated into the SAP HANA system (see section 3.4 below) without data with special tools and then the data will be selected (Analytic steps, para. 23). Although the graphical user interface (GUI) at the beginning could be similar as the existing SAP ERP to a certain extent, there will be also several Fiori apps available to process the different business scenarios (see section 3.5 below). Currently Leica Geosystems has a SAP training and support centre in Barcelona with six employees. They train the SAP users traditionally with documentations either using Microsoft Word or PowerPoint documents. In this sense Leica Geosystems AG has a huge effort ahead in order to train employees when SAP S/4 HANA Fiori apps get implemented. This effort will require a big human capital, as well as corresponding resources and costs. Also, the trainings need to happen quite quickly, and the effectiveness of the trainings need to be ensured. The effectiveness is crucial to avoid a lot of support that would limit the operability of the company. As a result, this addresses the question about how these trainings with Fiori apps could be created in a way that they are effective in terms that users retain the knowledge after passing first time, and also in a way that the company optimises the utilisation of SAP training resources. When addressing this question, it is important to think also in sustainable learning solutions. The sustainability refers to the SAP training of customised and adapted SAP business processes and workflows based on customers' requirements now and in the future. Although it is the interest of the company Leica Geosystems AG to implement standard SAP business processes as much as possible, in most cases SAP business processes need to be customised internally. With other words it can be said that standard or existing SAP processes are not something rigid or unchangeable, but always subjected to improvements. This implies that the customisation and adaptation of SAP business processes, as well as subsequent trainings need also to be effective and efficient.

Travelling has become due to the economic inflation in 2021 and 2022 quite expensive, and it is expected by the company Leica Geosystems that online trainings are more and more available worldwide. Because the training and support teams sit in Spain, online trainings are quite difficult due to different world time zones and the current limited number of available SAP trainers. Additionally, these SAP trainers do not only train, but also have to support the entire organisation worldwide on daily basis. In this sense e-learning is a quite interesting and attractive option because it enables online trainings 24 hours a day, 7 days a week, independently of the time zone. There are many e-learning tools in the market, however most of them are seen more as tools, where videos and files can be imported, and less as a e-learning artefact with a learning or didactic method validated rigorously and scientifically in the background specifically for SAP S/4 teaching purposes. SAP S/4 HANA provides in general e-learning tools, but they are based on SAP standard processes and many times this does not satisfy companies´ requirements. Also, there has been some e-learning artefacts created with a scientific method in the background. However, these studies carried out at educational institutions are based on standard SAP processes only, or the studies validated in companies have a very small sample size in terms of 16 (Conroy, 2012, p. 65) or 52 users (Deranek, K., McLeod, A., Schmidt, E., 2019, p. 376). The e-learning artefact for SAP S/4 HANA Fiori apps should be available worldwide and the language should be English. English is Leica Geosystems AG company's language. This implies that the internal communication per phone or email, as well as all internal documents and trainings are made only in English. Job descriptions of the company also remark the importance of speaking English fluently. If English would not be a requirement for the employees, the translation of documents and trainings in several languages could cause cost inefficiencies through the entire organisation worldwide. The e-learning artefact should also be inexpensive and easy to use. It should be able to use Sharable Content Object Reference Model (SCORM) files, as well as standard multimedia video formats like e.g., mp4. The e-learning artefact should be integrated into the company's security systems and be reachable only through the company's virtual private network (VPN). Additionally, users should enter a username and a password to enter into the e-learning artefact. Further, it should be accessible through personal computers (PC), tablets and smartphones. This is quite important because the learning environment of the users should be flexible. Flexibility means that they can either learn during their work time at the company, during home office at home, or during a business travel with the train, plane, etc. Flexibility gives employees the chance to look for a comfortable place when doing the course or alternatively to utilise and optimise their time schedules. It is the intention of the company Leica Geosystems AG to use the e-learning artefact for several SAP S/4 HANA business processes and workflows. In this sense the next chapter provides an overview of SAP S/4 HANA business processes with examples related to the business environment of the company Leica Geosystems AG.

3.3   Introduction SAP S/4 HANA

The new SAP Business Suite 4 SAP HANA, so-called SAP S/4 HANA (High Performance Analytical Appliance), is an ERP (Enterprise Resource Planning) System, as well the new version of the current SAP ECC (ERP Central Component) system. In comparison to SAP ECC, S/4 HANA is based on a new SAP HANA database and aims to support companies in a new digital transformation triggered by technology innovation, new customers' requirements, and subsequent

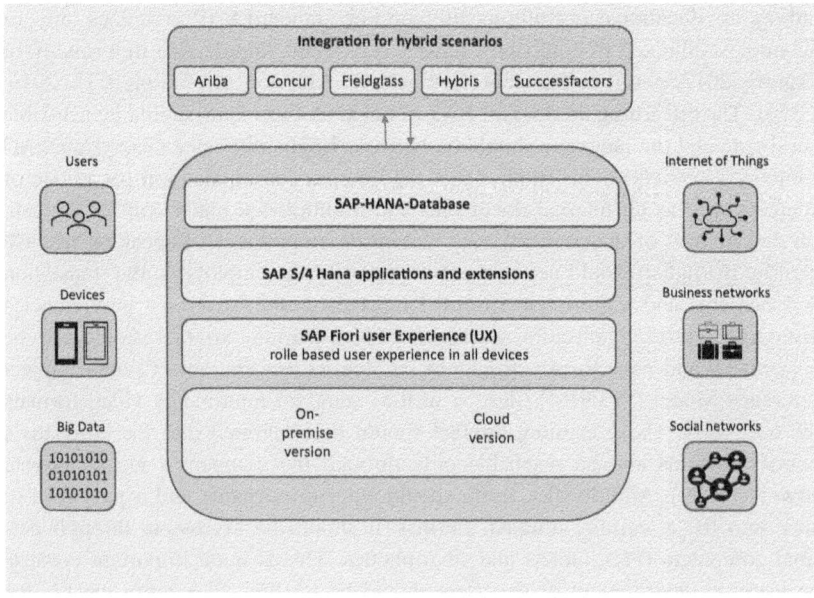

Figure 3.1 Modern requirements and SAP S/4 HANA (adapted from Koglin, 2016, p. 41)

new business processes (Koglin, 2016, p. 13). Figure 3.1 gives an overview about the core of SAP S/4 HANA, and the areas where modern requirements are needed in terms of Internet of Things (IoT), business networks, social networks, users, devices and big data. It also shows an integration with other hybrid scenarios in the sense of systems integration with Ariba®, Concur®, etc. The core of SAP S/4 HANA is represented by the SAP HANA in-memory database, the SAP S/4 HANA applications and extensions, the SAP Fiori user experience, and the deployment options (on-premises version and cloud version).

3.4   The core of SAP S/4 HANA and fulfilment of modern requirements

The company SAP AG started many years ago with an existing in-memory technology machine called TREX that saves flat files in a database (Berg, & Penny, 2013, p. 36). Additionally, SAP AG also developed a SAP-live cache technology that enables the processing of big data thanks to an object-orientated database technology based on a MaxDB relational SAP database (Berg, & Penny, 2013, p. 36). This SAP-live cache technology is especially beneficial when it comes to accelerate complex and time-consuming SAP applications inside supply chain management, or within advance planner and optimiser, as well as SAP BW Accelerator (Berg, & Penny, 2013, p. 36). Further, the SAP AG developed a complete in-memory solution called SAP HANA database thanks also to the development of the hardware technology. At the same time, with this development, SAP AG could reduce costs in the production of the databases and increase their memory capacity (Berg, & Penny, 2013, p. 37). SAP HANA enables the processing of big data in real-time thanks to the utilisation of cells and columns from hardware and software, as well as utilising object-based database technologies (Koglin, 2016, p. 54). This represents a major development, because in current SAP ERP systems data is saved in databases utilising cells only (Koglin, 2016, p. 54). In the case of the SAP HANA database, the additional column-based approach enables an acceleration when the user accesses the database (Koglin, 2016, p. 54). SAP HANA also saves and processes data in the main memory, causing a minimal latency when processing data (Koglin, 2016, p. 54), This is also possible, because SAP HANA can compress very big data volumes in little time (Berg, & Penny, 2013, p. 37). The SAP S/4 HANA applications refer to the SAP S/4 HANA Enterprise Management or so-called Line of Business (LoB) products, as well as to the compatibility packs (Destradi et al., 2019, p. 40). Several LoBs are supported by SAP S/4 HANA (see Table 3.1). A compatibility pack is an SAP business suite application that enables the usage of an defined SAP ERP application during a limited period of time and can be installed together with SAP S/4 HANA (Destradi et al., 2019, p. 43). These compatibility packs cover the areas of asset management (e.g., maintenance plans), controlling (e.g., amortisation and profit-centre), personnel management (e.g., time management), manufacturing (e.g., sales operating planning), R&D (e.g., project management), sales (e.g., bonus), sourcing and procurement (e.g., discounts), and supply chain (e.g., warehouse management). Thanks to the compatibility views it is possible to access the same SAP tables, while continuing to use the SAP S/4 HANA system (Destradi, p. 467). Compatibility packs are important, because the new SAP S/4 HANA will allow to work to a certain extent in the same way, as with SAP ERP was the case. This will be the situation for certain applications at least during a limited period. One example of these compatibility packs is the module of SAP Customer Service (CS) module. SAP S/4 HANA does not include so far any SAP CS module, however the compatibility packs will support it through the function service agreement and warranty management of these areas. In this way employees can keep creating service notifications and service orders in the system, book hours and materials, create debit notes and invoices. The applications of SAP S/4 HANA consist also of the so-called LoBs. These LoBs refer basically to different business processes in an organisation, as well as areas related to material master data and asset management data. There are two versions of SAP S/4 HANA, an on-premises version, and a cloud version. While the on-premise version is installed in a server within a company's own network and IT infrastructure, the cloud version is accessed via the outsourcing of services like storage location, computing power or application software through the internet (Destradi et al., 2019, p. 69).

SAP S/4 HANA also represents an innovation in terms of GUI with a new user experience (UX) and strategy. In this sense SAP S/4 HANA distinguish three phases in the UX strategy: SAPUI5, SAP Fiori and SAP Personas 3.0 (Destradi et al., 2019, p. 81). It is the idea of SAPUI5 to provide applications based on tasks, roles, as well as make them available in mobile devices like tablets or smartphones, as well as independently of operation systems or web browsers (Destradi et al., 2019, p. 83). Further, SAPUI5 enables the creation of transactions, reports, and dashboards, and is based on the programming language Hypertext Markup Language (HTML5), Cascading Style Sheets (CSS), Javascript and jQuery (Destradi et al., 2019, p. 83). The key idea of SAP Fiori is that users can work based on roles and tasks through so-called Fiori apps (Koglin, 2016, p. 85). There are three types of Fiori apps: transactional, fact sheet and analytical Fiori apps (Destradi et al., 2029, p. 89). With transactional apps users can carry

Table 3.1   SAP Enterprise Management or LoB (Destradi et al, 2019, pp. 41-42).

LoB	Compatibility areas
Data protection	Block and delete employees related data.
Database and data management	Master data management.
Asset management	Maintenance operations, resource scheduling, geographical enablement framework for asset management, environment, health, and safety.
Commerce	Subscription billing and revenue management.
Finance	Financial planning and analysis, accounting and financial close, treasury management, commodity risk management, financial operations, contract accounting, governance, risks, compliance for finance, real estate management.
Human resources	Organisational management, timesheet, SuccessFactors employee central connectivity.
Manufacturing	Manufacturing engineering and process planning, manufacturing for production engineering and operations, production planning, manufacturing execution for process industries, outsourced manufacturing, quality management, maintenance, repair.
R&D / Engineering	Enterprise portfolio, project management, product development foundation, product lifecycle management, product compliance.
Sales	Order and contract management, commodity sales, incentive, and commission management.
Service	Service agreement and warranty management.
Sourcing and procurement	Supplier management, sourcing, purchase contract management, central procurement, guided buying integration for central procurement, operational procurement, procurement analysis, invoice management.
Supply chain	Inventory, warehousing, transportation, order promising.
Analytics technology	Process performance monitoring, query designer, predictive model analytics.

out different tasks through business process steps in the context of workflows. For example, in the sales process there are different transactions for different tasks. These transactions refer to the steps like creation of a sales order, creation of an outbound delivery, pick outbound delivery, and creation of an invoice. For every single step of this sales process there is a transactional app. This transactional app can exist and be in an standard way available in SAP S/4 HANA, it can be customised according to the business process requirements, or it can be new programmed. Additionally, to the transactional apps, SAP S/4 also offers the Fact Sheet apps. The main objective of the Fact Sheet apps is to provide information about master data like i.e., articles or vendors, but also about relevant business information like stocks availability (Destradi et al., 2019, p. 90). Further analytical apps are used for the representation of key performance indicators, for queries or for reporting purposes (Destradi et al., 2019, p. 90). For example, with analytical apps it is possible to run a report about the orders in hand (OiH) received. These are special and important reports for companies e.g., to identify, if the budget in a specific month or quarter is going to be achieved. These types of reports plus the benefit of real time data empowers companies to make quicker decisions and take short-term measures to ensure that financial objectives can be achieved.

The digital transformation of SAP S/4 HANA aims to fulfil modern requirements in environments like social media, IoT or business mobility (Koglin, 2016, p. 24). In the case of Leica Geosystems AG, the integration of social media like e.g., Twitter, Facebook or Linkedin in SAP, is crucial. For example, companies use such social media for searching new employees, for marketing campaigns, or for getting feedback about new product or services. IoT at the same plays an important role when it comes to integrate the data from hardware equipment like total stations (electronic theodolites), laser scanners, or global reference positioning networks into the cloud. In this sense data integration of customers' hardware equipment can be used for hardware preventive maintenance activities. Mobility solutions have also become more and more frequent in companies' business environments. Thus, e.g., installers or service technicians on the field need to report, with field service solutions embedded in smartphones or tables, labour hours as well as components needed for a service at customers' places. Additionally, these installers or service technicians need to utilise current google maps to optimise the route for visiting customers.

Several factors influence a modern IT-System in the context of new digital requirements: flexibility, big data management, usability, and real time analysis (Koglin, 2016, p. 31). SAP interfaces are able to interact with other technology systems, like i.e., the Customer Relationship management (CRM) system called Salesforce.com or software license servers like FlexNet or Thales. This implies a high level of interconnectivity through the programming of interfaces among systems. Also, IoT solutions generate big amounts of data. The management of this big data from e.g., hardware systems with relevant information like hours of equipment utilisation, oil pressure, etc., means at the same time new technical challenges in terms of having the needed IT storage capacity and corresponding IT infrastructure.

3.5   Integration of Business Processes and New Features in SAP S/4 HANA

This chapter gives an overview of the different business processes that are supported by SAP S/4 HANA as well as the new features in this new version in comparison to SAP ERP. The concept of integration means the modelling, the data exchange between applications and systems, the configuration of applications for specific business processes, and the automatisation of business processes (SAP, 2014, p. 15). SAP S/4 HANA, as an integrated system, covers generally all business processes of a company. These business processes are related to Idea to Market, Source to Pay, Plan to Fulfil, Lead to Cash, Recruit to Retire, Acquire to Decommission, Governance and Finance (Sarferaz, 2022, p. 51). This chapter describes these business processes from the perspective of SAP S/4 HANA, but also their meaning in the business context of the company Leica Geosystems AG.

3.5.1   Process of Idea to Market

The process of Idea to Market covers the process from the idea development of a product until the design of products and services. The Idea to Market process contains the processes Plan to Optimise Products and Services, Idea to Requirement, Design to Release (see Figure 3.2), Product/Services to Market and Manage Product / Services (see Figure 3.3). The process *Plan to Optimise Products / Services* takes care of the management of the product and services strategies, of the products portfolio, as well as the corresponding needed portfolio investments. This process has the analogy with a product development funnel, where several products and services ideas are evaluated to decide which products and services should be optimised. The process *Idea to Requirement* comprehends the Ideation management and the product, as well as services requirements. The Ideation management contains the subprocesses perform discovery research, generate new concepts as well analyse new ideas and requirements. The deliverables of the process *Idea to Requirement* are a vision of a solution, a project scope, and a first risk assessment for potential technical, commercial and management risks. The project scope includes the situation before and after of a planned project, the content and non-contents of the idea, as well as goals and non-goals. Additionally, it includes the project environment in terms of stakeholders. The business plan gives an overview about the market and competitor analysis, a SWOT (Strengths, Weaknesses, Opportunities and Threats) analysis, a marketing mix including product, price, promotion and place policy, and a financial analysis. The product and service requirement demands a system idea document, where use cases, as well as the product and service requirements need to be defined. After the Ideation management, the technical and product requirements also help to define the architecture and detail concepts of product and services.

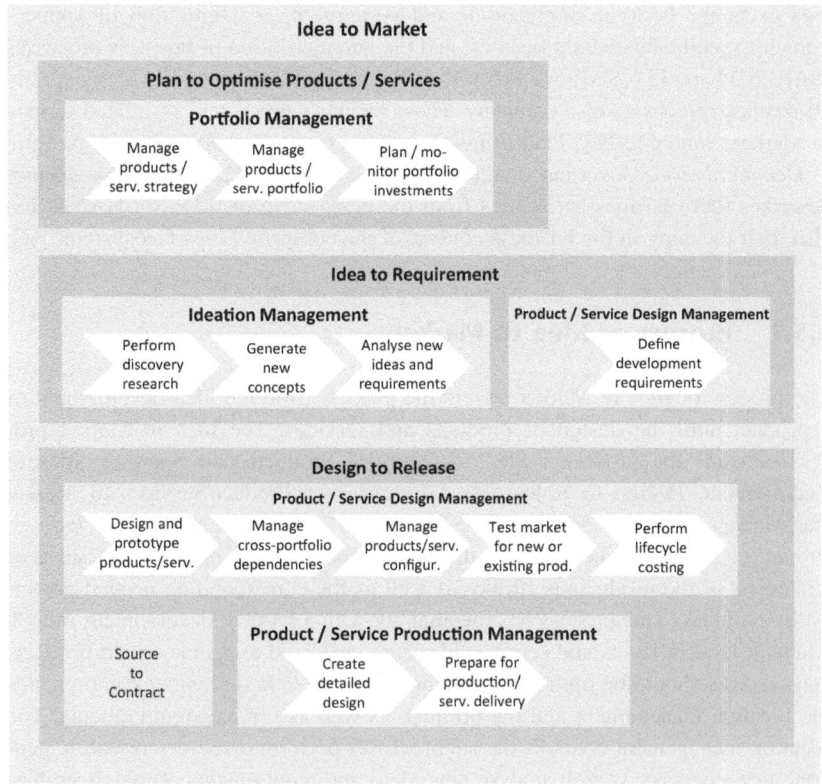

Figure 3.2 SAP S/4 HANA Idea to Market process (Sarferaz, 2022, p.102)

These requirements also include the future test plans as well as usability evaluation in terms of how potential customers would use a product to achieve their own goals. The process *Design to Release* covers two major subprocesses, the product and service design management, and the production management. The design management defines the needed processes required from prototype creation until the performance of lifecycle costing. The requirements in the previous process are the basement for the design of a prototype. At this stage the product or service manager defines the needed product variants like sales variants and upgrades in a product data sheet. The product data sheet is the template for the product data management, where the material master of BOM (Bill of Materials) and the computer aided designs (CAD) are defined for a later production. The tests of the prototype ensure the validation of the product requirements and the fulfilment of the users' cases. The tests are normally carried out in the market with two to three key customers. After the tests, the corresponding validation and the improvements have been carried out, it is important to ensure that planned production costs like e.g., assembly overhead costs and material costs are covered. This happens independently if the production is carried out in the own company, or alternatively if it is outsourced through contracts to third parties. The product and service production management relates to the creation of a detailed design and the preparation for production. While the detailed design is the result to the last improvements made after the test phase, the production is related to the estimation of the BOM (Saferaz, 2022, p. 106) needed for the estimated quantities that will be absorbed over a timeline by the market.

Additional processes of the Idea to Market are represented through *Product/Services to Market and Manage Products/Services* (see Figure 3.3). The *Product/Services to Market process* includes market offering, the development and management of pricing, and the measurement of product satisfaction. At this stage is where the product announcement and product release into the market takes place. The product announcement requires, among others, the preparation of all marketing digital materials, the trainings of the sales representatives, and the distribution of demo instruments to sales representatives. The process *Manage Product/Services* include the subprocesses of Product/Service Lifecycle Management, the Product Compliance Management, and the Product/Service Development Collaboration. The *Product/Service Lifecycle Management* relates to the projects during the lifecycle of the product and services. For example, the after sales processes play here an important role in terms of maintenance contracts for hardware of services, the calibration of services, or even the upgrade of onboard software. The subprocess *Product Compliance Management* investigates the regulatory requirement like e.g., the ones related to the transport of dangerous goods in terms of batteries (Saferaz, 2022, p. 112). The *Product/Service Development Collaboration* is basically a platform upon individual companies can acquire knowledge about customers' additional feedback that results from the utilisation experience. This utilisation experience is a valuable know-how for future innovation projects.

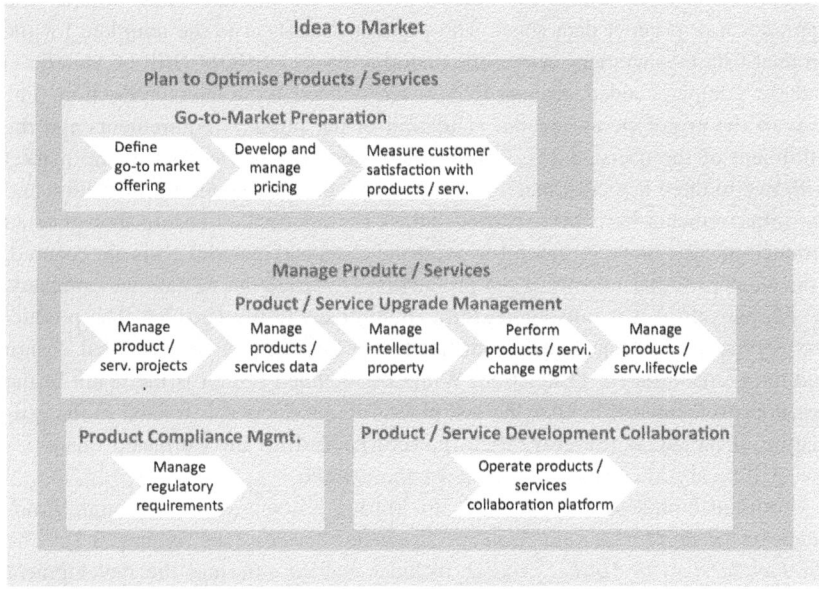

Figure 3.3 SAP S/4 HANA Idea to Market process (Sarferaz, 2022, p. 110)

3.5.2 Process of Source to Pay

The processes of Source to Pay include the processes of Plan to Optimise Sourcing and Procurement, Source to Contract, Procure to Receipt, Request To Resolution, Invoice to Pay and Manage Suppliers and Collaboration (see Figures 3.4 and 3.5). *The process Plan to Optimise Sourcing and Procurement* aims primarily the analysis of potential suppliers and provides an overview of the required and needed purchases (Sarferaz, 2022, p. 122). The idea of this analysis is to try to get the best proposals (Sarferaz, 2022, p. 122). The best proposal is determined by several factors like costs, quality, failure rate, local or global sourcing possibilities, terms of payment, etc. Once the analysis is carried out, the Source to Contract Process comes into play. The RFX refers to request for information, request for quote and request for proposal (Sarferaz, 2022, p. 122). The result of the RFX should end up with the selection of the best proposals with the aim to enter in a so-called negotiation phase with the providers. The negotiation phase should ensure a long-term partnership, but without exclusivity. For example, a high-tech company should ensure in its procurement strategy, that more than one supplier can deliver the needed parts. This especially becomes crucial in difficult social and economic situations like the past covid infection in 2020, or the subsequent crisis related to the delivery transportation. *Procure to Receipt* represents the operational process that consists of managing the purchase requisitions and corresponding purchasing orders with the external providers. In this process it is important that the contract specifications and agreements are met in terms of availability to promise (ATP), keeping the agreed prices, etc. The dock and yard logistics refer to the co-ordination of the loading bays as well as the traffic given in certain premises (Sarferaz, 2022, p. 123). The last subprocess of the *Procure to Receipt* is linked to the post goods issue of products and services.

While the products are posted into the company stocks, the services are normally consumed. However, many times, it could happen that products and services are resold. This is the case e.g., when a company buys software licenses with the last maintenance updates, build this software into companies´ devices and resell it to their own customers. Another relevant process of the Source to Pay is the *Request to Resolution* (see Figure 3.5). The *Request to Resolution* process is about the management of supplier claims and returns management. If suppliers do not meet the quality agreements, product and services will be returned, or warranty claims be placed. This requires the establishment of a standard process that simplifies the return of goods and the creation of credit notes or the replacement of the same goods. Also, the quality agreements should specify if the products and services should be swapped or if a credit note should be issue. The process Invoice to Pay relates to the management of payments from the suppliers. According to the contract agreements, payments are made. Payments are done currently more and more through e-invoice directly from SAP.

Finally, the process Manage Suppliers and Collaboration covers basically the supplier management processes related to the management of the suppliers in terms of certification and validation, evaluation of the supplier performance and the creation of a collaboration platform to operate the daily activities with the supplier.

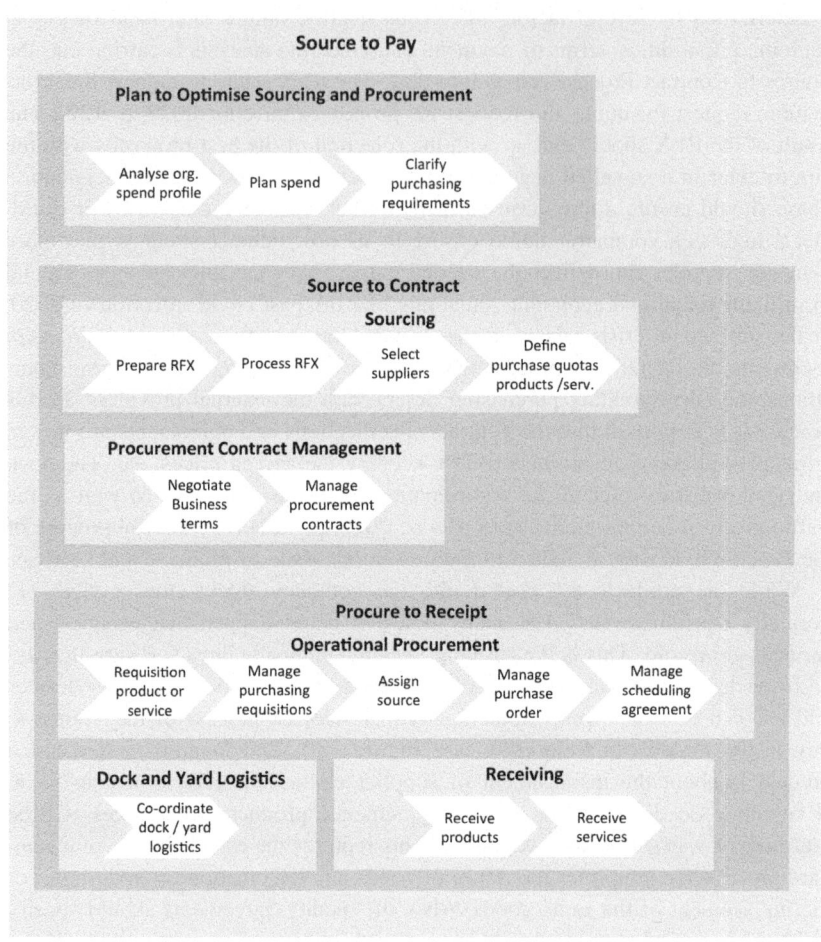

Figure 3.4   SAP S/4 HANA Source To Pay process (Sarferaz, 2022, p.120).

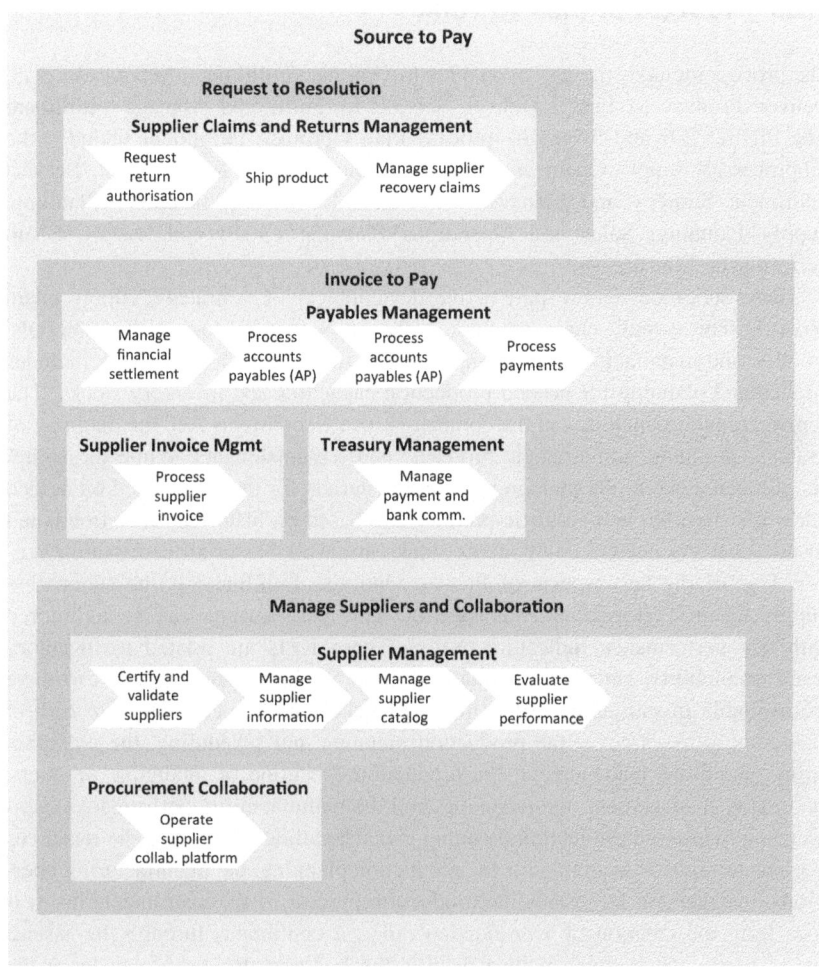

Figure 3.5 SAP S/4 HANA Source to Pay process (Sarferaz, 2022, p. 121)

3.5.3   Process of Plan to Fulfil

The process includes the processes Plan to Optimise Fulfilment, Make to Inspect, Deliver Product to Fulfil, Deliver Service to Fulfil and Manage Fulfilment (see Figures 3.6 and 3.7). The process Plan Optimise Fulfilment includes the subprocesses Supply Chain and Manufacturing Strategy Management, Service Fulfilment Strategy and Management, Demand Planning, Inventory Planning, Supply Planning, Sales and Operations Planning (SOP) and Supply Chain Performance Management.

These subprocesses are part of the development of a strategic supply chain network. This supply chain network is the basis for developing the materials, product, and manufacturing strategies. Further, the supply chain network includes the demand planning for needed production capacities and inventory stocks. The demand planning includes, at the company Leica Geosystems AG, the planning of strategic products, non-strategic products, and replenishments. While the strategic products refer to the planning relevant materials for the SOP based on agreed sales forecasts, the non-strategic products are based on historical sales trend, and ensures that the needed components stocks are available in all distribution centres. One of the key aspects of Plan to Optimise Fulfilment is the subprocess Supply Chain Performance Management. This performance can be monitored with key performance indicators (KPIs). These KPIs are related to inventory reach, availability, and forecast quality. These are important indicators to meet the available quantities according to the demand. The process *Make to Inspect* main subprocesses cover the production planning and scheduling, the manufacturing operations management, the production execution of intangible products, the quality management, the receiving, and the manufacturing performance management. While the production planning and scheduling targets that the resources (e.g., materials) are available for the production planning, the manufacturing operations management represents the production process of the instruments itself. It goes from the creation of a production order, it continuous through the assembly of parts, and it finish with a quality check. Later the goods can leave the production facility. The production execution of intangible products refers to the creation of services (Sarferaz, 2022, p. 142). In this sense, onboard software for hardware instruments represents an intangible product whose license need to be created in a license server. Further quality testing is a needed step to avoid future complaints from customers which could have a negative impact in a company's businesses. Through the Manufacturing Performance Management, the production processes can be analysed and improved in the context of continuous improvement processes. In this way production efficiency improvements, that drive the

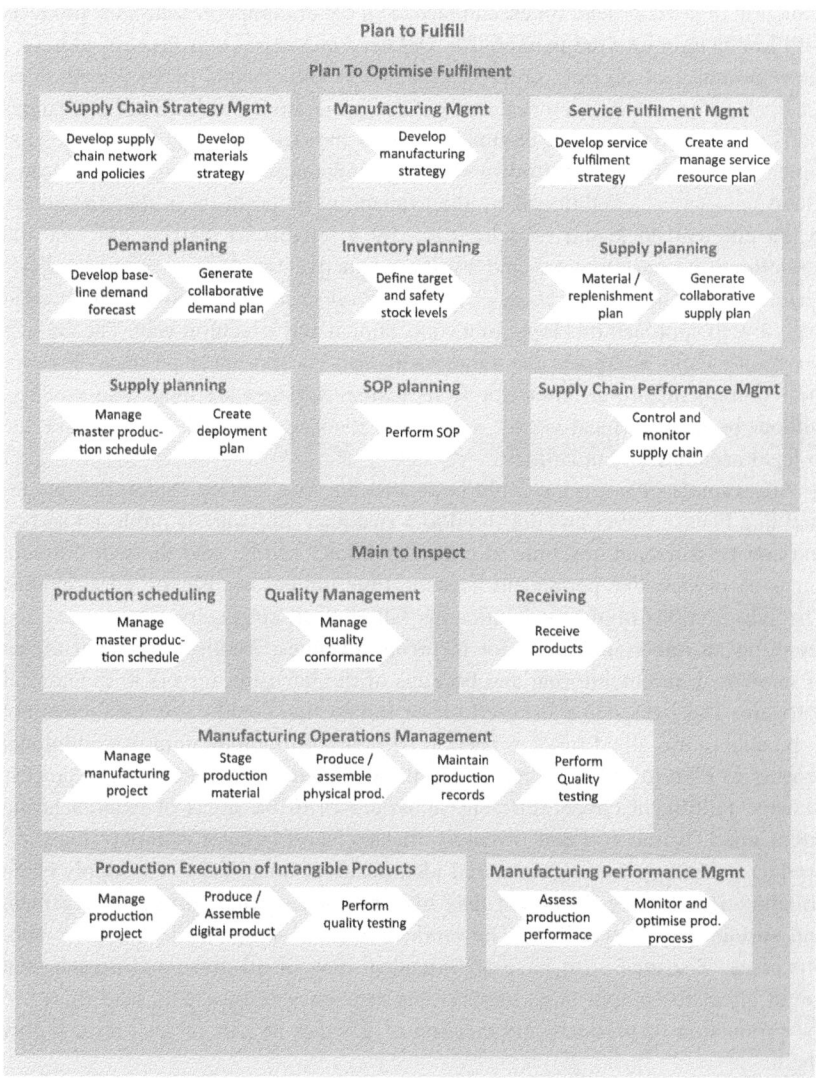

Figure 3.6   SAP S/4 HANA Plan to Fulfil process (Sarferaz, 2022, p. 144)

reduction of costs or lead times, can provide a better competitiveness for products with low margin into the market. The *Deliver Product to Fulfil* Process addresses the importance of on time order fulfilment for selling units of an organisation, for dealers, or end customers directly. Customers from international companies like Leica Geosystems AG commits itself to the delivery of goods to the customer on time. This is by itself a challenge influenced due to different reasons. These reasons are conditioned by the availability of products in the warehouse, by an efficient dock and yard logistics, by a reliable as well as efficient and cost-effective transportation, and by the pro-active monitoring of the logistics performance. The subprocesses of *Deliver Product to Fulfil* provide the needed overview to approach this topic in a conceptional and structural way. The *Deliver Service to Fulfil* addresses the same challenges as it were a product, however the requirements are different due to its nature. Services are intangible products without booking in local stocks, except for countries like India and Russia due to legal accounting standards.

An example of a service could be a software as a service (SaaS) that is created in a license server and delivered to a customer. Software is produced ad-hoc and can be delivered any time to customers. SaaS can be sold through different licensing models like permanent licenses, time limited licenses or subscriptions. The subscriptions imply the continuous billing until customers cancel them and represent an important source for recurring revenues. Another important aspect in services in the monitoring and backups of the licensing servers in the case of software. The risks due a licensed server failure that could cause customers not to be able to use cloud services need to be reduced to a minimum. An additional process to Plan to Fulfil is represented through Manage Fulfilment. The process Manage Fulfilment covers different subprocesses in the areas of manufacturing engineering, warehouse and inventory management, circular economy logistics, product genealogy, logistics material identification, track and trace, supply chain collaboration, service fulfilment data management, service partner management and sustainability operations. The production processes consider all these subprocesses. The idea is to have an extended view of all areas and aspects that might affect the output (e.g., quality) and efficiency results (e.g., lead times) in the production of products and creation of services as part of the entire supply chain.

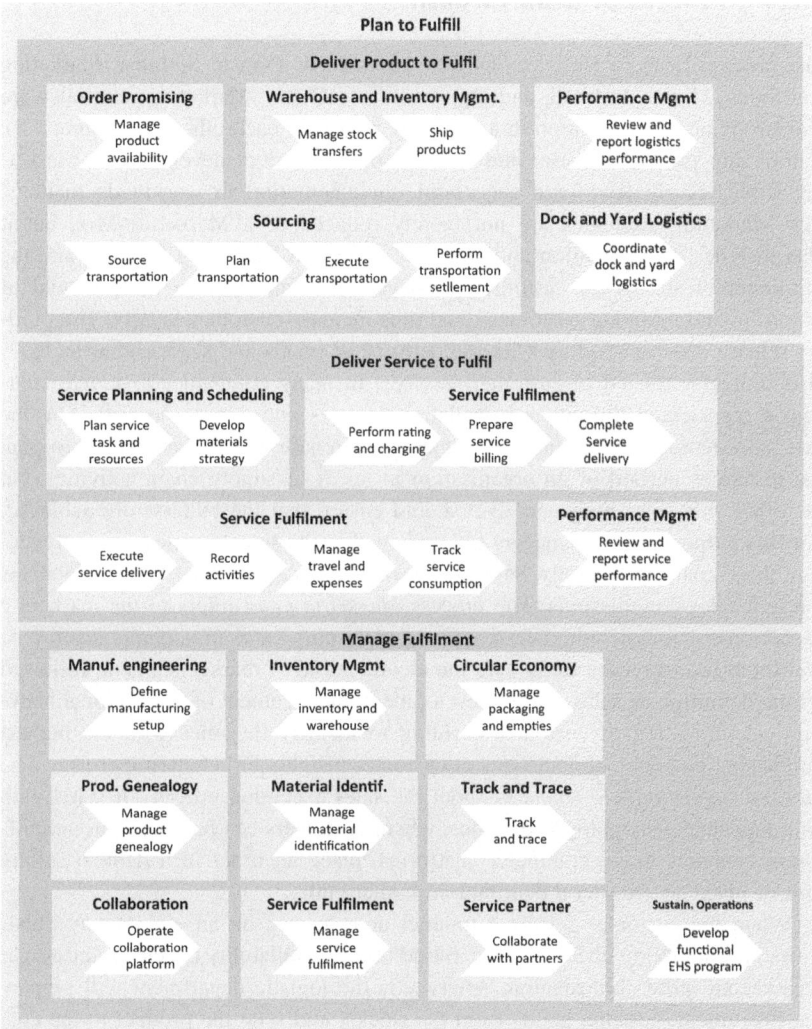

Figure 3.7   SAP S/4 HANA Plan-to-Fulfil process (Sarferaz, 2022, p. 145)

3.5.4   Process of Lead to Cash

The process Lead to Cash covers the processes of Plan to optimise Marketing and Sales, Market to Lead, and Opportunity to Quote. Marketing and Sales are two strong areas of attention in a company related to each other (see Figure 3.8). Before any product release, and as part of the product development, Marketing strategies are key for a successful implementation in worldwide markets. The Marketing strategies are not merely reduced to a Marketing-Mix, but it also includes the estimation and approval of needed Marketing budgets, and the fulfilment of successful customer journeys. This can enhance the potential of acquiring and retaining customers, and thus increase customer loyalty. The product release does not end up with the delivery of goods and services but includes also strategies related to customer services. In many companies a service strategy is represented through a so-called after sales and service strategy. On one side sales representatives and sales managers prepare every year budgets to plan the resources needed of an organisation in terms of supply chain activities. On the other side sales managers also should ensure that the budgets are achieved. For this purpose, sales managers set goals within the sales organisation and measure these goals accordingly. Sales managers also incentivise sales representatives through sales commissions. The process *Market to Lead* refers to the marketing execution containing subprocesses whose main purpose is to generate leads. The *Market to Lead* process starts with the identification of market segment, followed by the definition of sales prices, development, management of promotional activities, identification of customer profiling to identify the ones with the highest purchasing potential, and analysis of customer insights for a better approach. The process *Opportunity to Quote* is about the sales execution process. It starts with an omnichannel shopping experience, where customers can purchase through different channels like e-commerce platforms, place an order in a defined selling unit or place an order at a distribution partner entity.

When the customer selects a product or a service on an e-Commerce platform, an availability check is done. Based on this availability check, the customer receives an order confirmation. Afterwards the logistic department will prepare the instrument, post the goods from the stocks, and send the product to the customer. Once the customer has received the products from a freight company, customer care will create an invoice. Important in the sales execution process though is that the customer journey and experience is easy. The more a customer can use an electronic commerce (e-commerce) platform in a satisfying way, the better. E-commerce platforms enables the potential reduction of costs in the sales process, as well as eventually a higher profitability. Other benefits of the

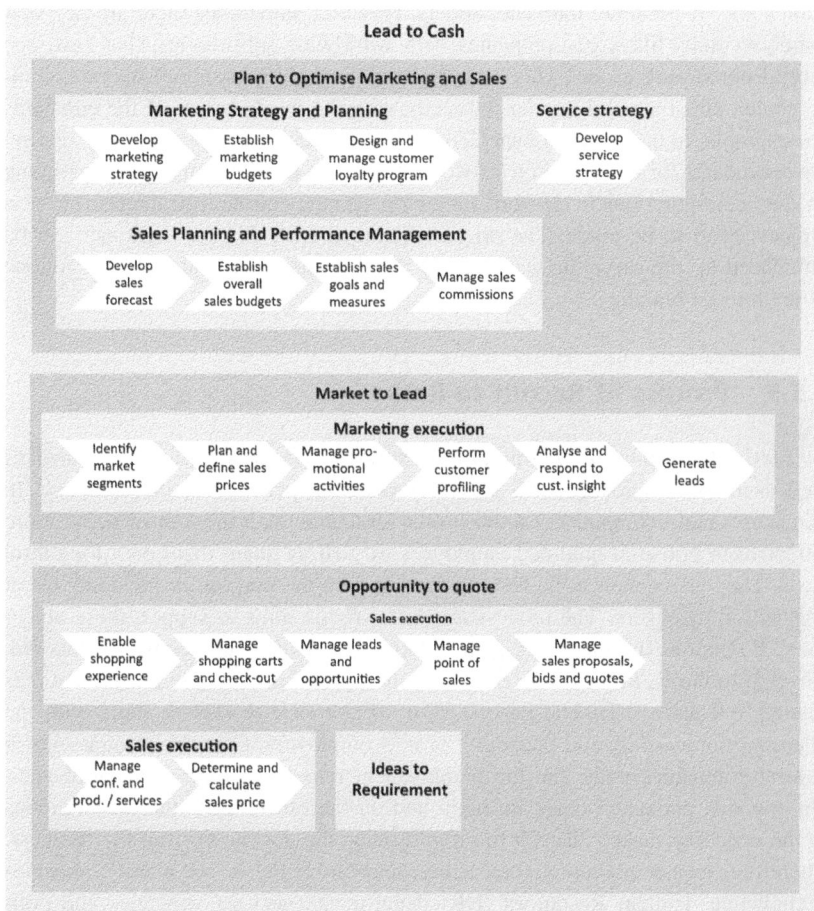

Figure 3.8 SAP S/4 HANA Lead-To-Cash process (Sarferaz, 2022, p. 162)

e-commerce approach are the hardly inexistence of discounts negotiations with customers, products and services can be sold to a better price, and the purchasing decision process is quicker. The *Opportunity to Quote* process also include the management of leads and sales opportunities, the point of sales, the receipt of proposals, bids, and quotes from key customer accounts. The sales team is responsible for managing the leads and opportunities that are later converted into quotes and, if possible, into sales orders. However sometimes there are key customer accounts like e.g., companies with worldwide subsidiaries, that have one central purchasing group. This central purchasing group wants to interact, communicate, negotiate and purchase directly with the headquarters of the company. For example, in the case of Leica Geosystems AG, customers want to buy surveying equipment for all their projects worldwide. This implies that certain surveying product configurations in terms of surveying accuracy of the instruments for these projects need to be made. The prices negotiated with this key customer can be influenced by the surveying accuracy of the instruments, but also by the volume of units to be bought.

3.5.5   Process of Recruit to Retire

Currently companies in central Europe struggles to get qualified personnel as well as retain employees with several years of experience at the company. In this sense many companies establish own strategies with the aim of being more attractive to potential qualified employees as well as retain them on a long-term basis. They also enhance the HR department with the purpose of including talent acquisition managers. The process *Recruit to Retire* aims to support these objectives. It contains the processes Plan to Optimise Workshop, Recruit to Onboard, Develop to Grow, Reward to Retain, and manage Workforce and Retirement (see Figures 3.9 and 3.12). The process *Plan to Optimise Workshop* starts with the planning, forecasting, and budgeting of new employees. Cost-driven strategies in several companies make the top management reluctant to hire new employees. On one side personnel costs are high, and on the other side future uncertainties in the economy makes difficult to prognosticate the development of the business. Therefore, proper planning, forecasting, and budgeting is for many companies a challenge. Human Resources (HR) departments also set strategies and policies, forecast resourcing needs, and apply for a workforce budget. Considering the organisational structure and main needs and priorities of business, the HR department plans an employee demand. The process *Recruit to Onboard* focusses on the operational activities when recruiting personnel. It begins with a corporate development branding followed by the approval of headcounts. The approval needs the agreement with the top management. Based on standard job descriptions HR publishes them in the company's website, but also in social networks. When candidates apply for the jobs, the HR department of Leica Geosystems AG initiates a screening and selection process based on the job description. Normally more than two interviews take place with the potential candidates. On one side interviews give an overview about applicants' experience and capabilities, as well as check the information provided in the resume of the curriculum vitae and application documents. But on the other side, interviews also are a good platform to provide potential candidates information about the company. Once a candidate has accepted a position, HR prepares the needed contracts as well as the purchase order for the equipment needed by the employee like e.g., personal computers, desk, etc. (Sarferaz, 2022, p. 183). As soon as the contract has been signed up, the recruit process finishes. The process of hiring employees is nowadays in companies an important topic due to lack of personnel, however retention of employees has an equal or even a higher priority. The motivation of employees is crucial to retain them on long-term basis. Here is where the process *Develop to grow* plays an important role. It starts with setting the goals and reviewing the performance. One time per year managers must carry out performance management meetings with their employees. Based on defined goals, their achievements are checked. Also, the managers protocol the answers to questions in terms of what did go right or wrong, where the employees see their highest skills, which skills need to be developed to perform the job better, or in which areas the employees would like to develop, etc. Also, managers identify which additional career and professional development needs employees might need to perform the current job better. Managers also prepare employees for future jobs based on specific development programs. HR departments normally count with development programs for the continuous professional education of the employees, as well as development initiatives that match the requirements of the yearly performance management. The yearly performance management help managers to plan employees' career and support them in that process. The process *Reward to Retain* (see Figure 3.10) aims primarily through incentives, compensation plan and benefits to keep the retention rate of employees high. The costs of losing experienced employees can hardly be measured, however they are in terms of business growth and innovation potential acknowledged by the management. In innovation, for example, the loss of experienced employees could have an impact in the release of new technology products.

A salary increase itself is a short-term measure for the motivation. However, there are other compensations with a more long-term view. This compensation could be e.g., private pension plans, monetary incentives based on the results of the company, or the possibility to do some home office work. Also, a work life balance could make companies attractive to existing employees. The process *Manage Workforce and Retirement* (see Figure 3.10) supports the subprocesses related to HR Administration, Workforce Experience Management, Time Management, Procurement Planning and Analytics, Travel & Expense Analytics as

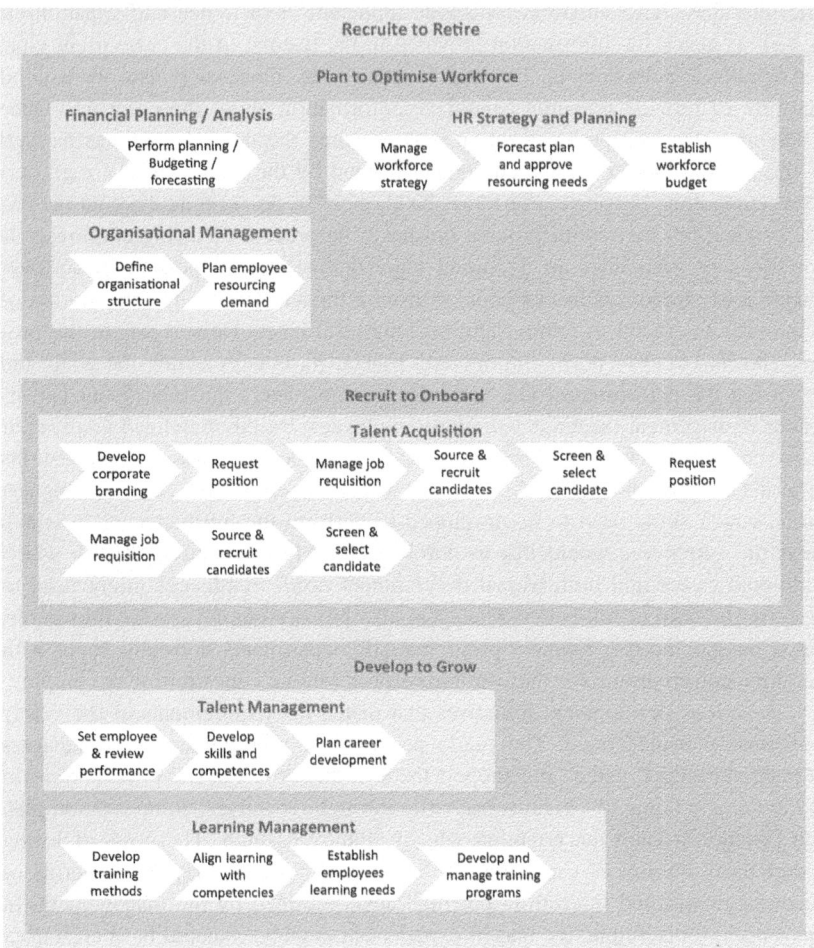

Figure 3.9   SAP S/4 HANA Recruit-To-Retire process (Sarferaz, 2022, p. 182)

well as Request, Travel & Expense Management, Payment and Reimbursements, and Invoice to Pay. The HR Administration takes care of operative activities related to maintain employee data (e.g., marital status), manage promotions in terms of new roles, relocate employees, manage employee requests like for example salaries increase, terminate employee contracts due to redundancies because

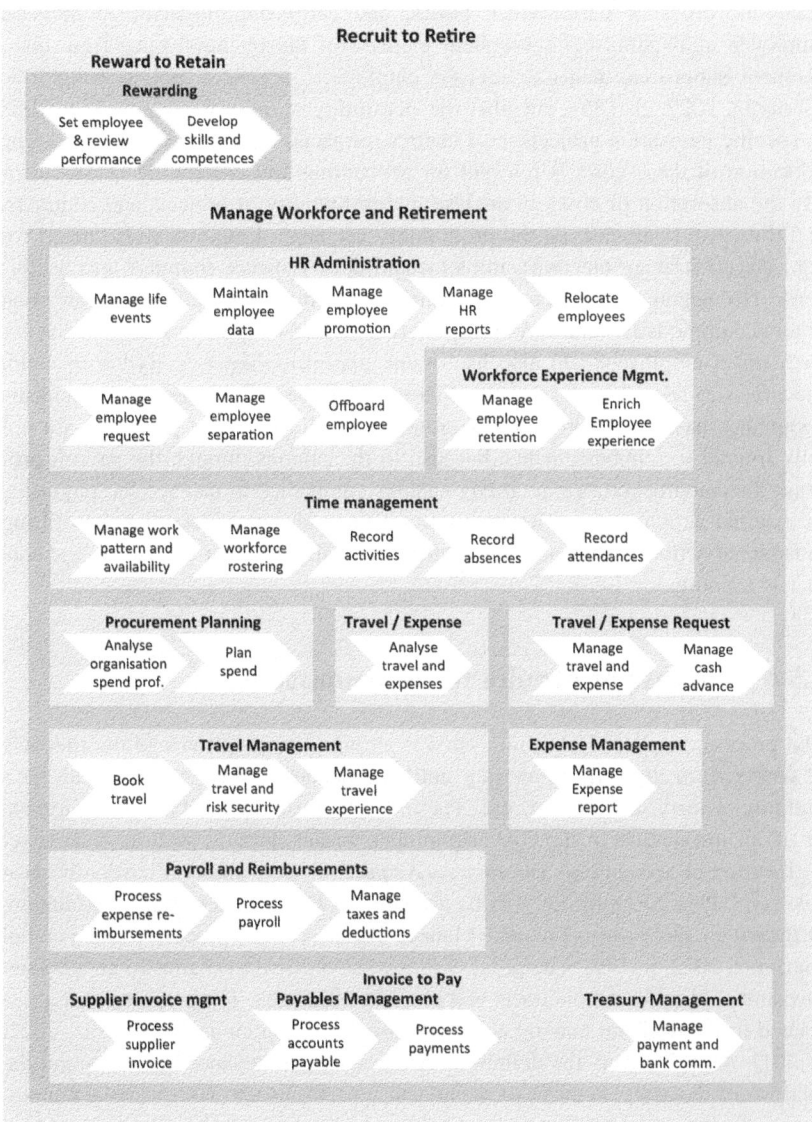

Figure 3.10   SAP S/4 HANA Recruit-to-Retire process (Sarferaz, 2022, p. 183)

economic crisis or performance issues, and carry out initiatives to increase employee motivation and subsequent increase of the retention rate. Time management subprocess helps to register employees' absences and working time (Sarferaz, 2022, p. 186), but also the possibility to allocate the working time to specific innovation projects, cost centres, projects, warranty accounts, etc. The allocation of these costs is relevant for accounting and controlling i.e., to identify the absorption of costs, to analyse the profitability of projects and contracts, and thus avoid variances in the profitability accounts. Employees also travel for e.g., due to strategy meetings, implementation of projects, trainings, etc. In this sense HR negotiates and prepares online platforms upon employees can book hotel accommodation and buy transportation tickets for planes, trains, etc. After each travel employees can also use online platforms like e.g., Rydoo to claim the expenses related to the travel activities. With the subprocess reimbursements companies pay the expenses to the employees. However, the payments are not only limited to reimbursements, but also to the salaries through the payroll process. The subprocess *Invoice to Pay* enables companies to manage the payments to external suppliers for HR activities; for example, payables related to hiring outsourced activities to head-hunters or to general consultant services in the area of HR trainings.

3.5.6   Process of Acquire to Decommission

The process Acquire to Decommission is a core process to manage the lifecycle of assets from the acquisition step until the decommission. An asset can be a building, a lorry, a computer, etc. These assets normally exist as an equipment or as an installation in the SAP equipment master data, as well as a fix asset registration in accounting. The process *Acquire to Decommission* covers the processes of Plan to Optimise Assets, Acquire to Onboard, Operate to Maintain, Offboard to Decommission, and Manage Assets. These processes run parallel (Sarferaz, 2022, p. 199). The process *Plan to Optimise Assets* aims to create asset investment plans based on assets and property strategies. The property strategy is related to the decision making about owning or leasing an asset (Sarferaz, 2022, p. 201). Also, it covers the definition of the asset maintenance strategies and the analysis of the assets in terms of maintenance performance. The technical maintenance of an asset can be done proactively or passively. The proactive maintenance approach means that SAP service orders are created in advance in the system for the following years to do maintenances on a specific date, while passive maintenance approach means that maintenances are done on demand. The proactive maintenance approach can e.g., be managed through SAP cockpits by a specific facility management team in a company. The *Acquire to Board* process comprehends the activities related to the asset acquisition, asset construction and asset commissioning (see Figure 3.11). The asset construction relates to situations, when a company develops and builds up itself an asset with own knowledge and technology. For example, a company can create a network of global positioning systems (GPS) reference stations in several countries and then sell the data of generated by these reference stations as a service. The whole reference network system contains GPS devices as well as the software that connects the GPS positions to these devices. Thus, customers can receive GPS positions of their GPS rovers thanks to the GPS corrections from the reference stations. Finally, before an asset is used productively, it needs to be onboarded (Sarferaz, 2022, p. 202). The process *Operate to Maintain* relates to the operative activities done to keep the assets in an active operating condition. In order to achieve this objective, a company needs to create maintenance plans and carry out standard service activities. The service frequency will depend on the asset type (e.g., if it is a building or a machine). The planning also requires the definition of resources in terms of materials and personnel. Later the maintenance can be performed. In some cases, the maintenance requires some refurbishment, where hardware and/or software will be updated to comply with the needed functionality. In our example of the reference networks, this would imply e.g., that the measurement engines of the GPS devices are exchanged to be able to be connected with a new software release for the network purposes. The process *Offboard to Decommission* covers the activities carried out when an asset is decommissioned (see Figure 3.12). It starts with the development of an exit strategy. This exit strategy will imply which assets will be decommissioned, and if they will be sold to third parties.

It could happen however that the assets are fully written off, their repair is not worth any more, and it needs to be scratched. In this case the proper environmental regulations for waste and hazardous goods need to be observed. During the whole process *Acquire to Decommission*, the assets need to be managed. The process *Manage Assets* supports this process during the entire lifecycle of the assets. It starts with the management of the assets master data. This enables the creation of assets in the financial accounts, but also as equipment in the equipment master data. The creation of an equipment is needed to process the maintenance works and record the technical history of an equipment in the SAP system. A risk analysis is also needed to evaluate, for example, the potential failures that assets might have, and to take preventive measures to avoid them. In this way companies can reduce the risks of operational costs that could incur due to potential technical failures of an equipment. In the case of the GPS reference stations, if one of them

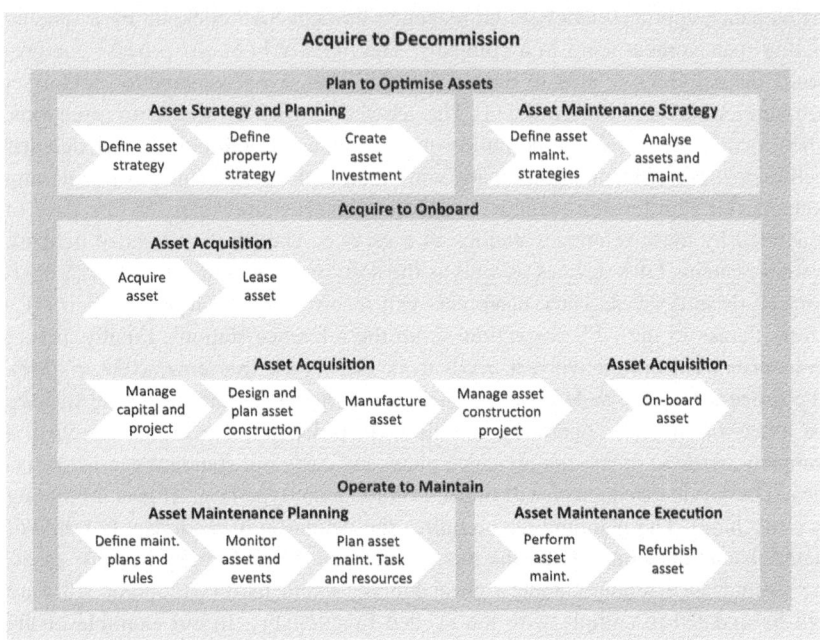

Figure 3.11   SAP S/4 HANA Acquire-to-Decommission process (Sarferaz, 2022, p. 200)

would failure, this could imply that customers in a geographical area would not receive a GPS reference signal. The management of assets also comprehends the sustainability operations in terms of environmental, health and safety regulations.

These regulations need to be observed e.g., when scrapping assets that are no more functioning or whose maintenance is not more worthy. Also, the employees who do the maintenance of assets need to observe the working health and safety regulations. The subprocess related to the Asset collaboration (see Figure 3.12) includes the performance of maintenance by own technicians in a company, but also by third parties, to whom the maintenance tasks are outsourced to.

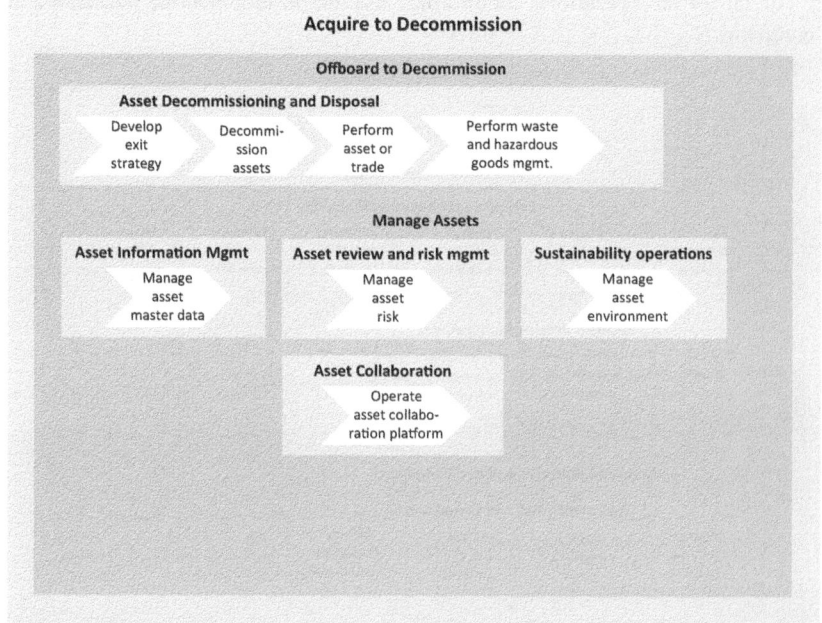

Figure 3.12   SAP S/4 HANA Acquire-to-Decommission process (Sarferaz, 2022, p. 201)

3.5.7   Process of Governance

The aim of Governance processes is to manage the company strategically and operatively (Sarferaz, 2022, p. 215). The Governance, as part of organisational shared services at corporate level in a company, include the processes of Plan to Optimise Enterprise, Manage Enterprise Risks and Compliance, Manage Identity and Access Governance, Manage Cybersecurity, data Protection and Data Privacy, Manage International Trade, Tax & Legal, Manage IT, and Manage Projects and Operations (see Figures 3.13 and 3.14). The process *Plan to Optimise Enterprise* includes the subprocesses related to corporate strategy and planning, as well as business process management. The *Plan to Optimise Enterprise* process comprehends the corporate strategy and planning, as well as the business process management. The corporate strategy also manages the business models, the operating model, the business information, and the branding strategy (see Figure 3.13). Further, the design, modelling, analysis and optimisation of business processes are crucial for the operations, monitoring, and the development of businesses in a company.

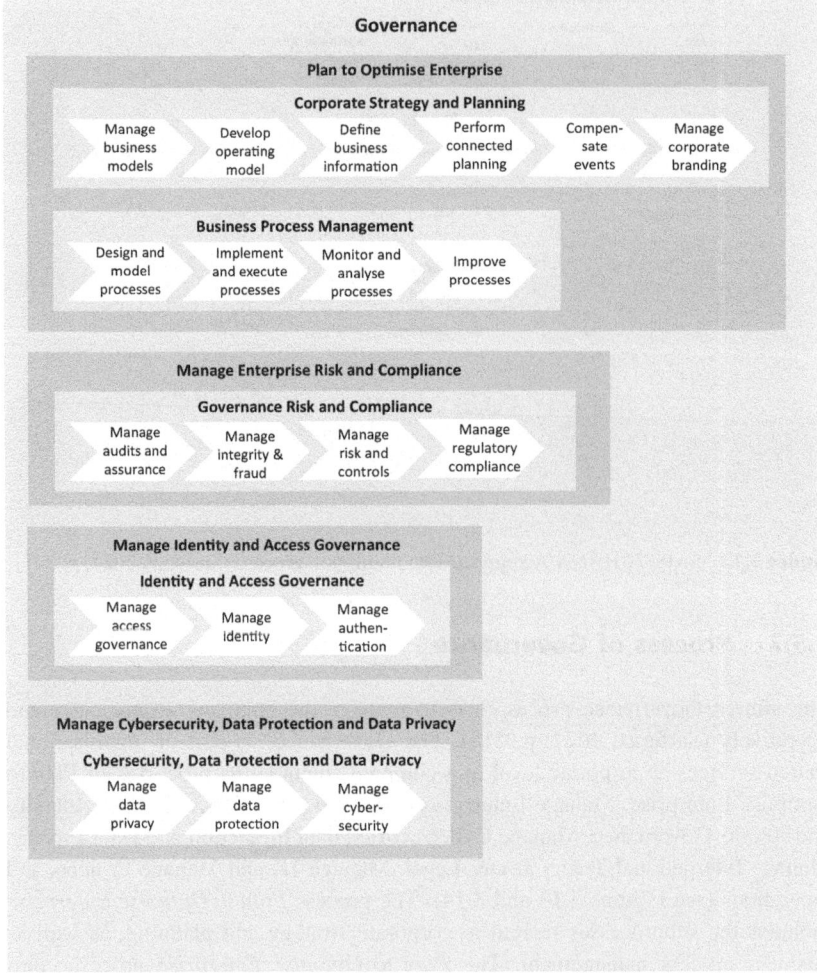

Figure 3.13 SAP S/4 HANA Governance process (Sarferaz, 2022, p. 216)

The process Manage Risks and Compliance relates to the observation of compliance regulations. Internal and external audits for quality and financial purposes belong to this risks area. In the case of compliance Leica Geosystems AG commits to the observation of high standards in ethics, as well as compliance to the fulfilment of all legal local requirements in countries, where the company operates (Schmitz, 2022, para. 1). The observations of compliances help to avoid fraud and guarantee integrity behaviour. They are needed for the compliance with legal, society and governmental good practices. It comprehends anti-corruption, competition, customs, export control, public contracting, and data protection (Schmitz, 2022, para. 1). The process *Identity and Access Governance* relates to the access management for different digital platforms like SAP S/4 HANA, for reports of the SAP business information warehouse, for innovation technology projects, etc. It also includes which roles give authorisation rights to further roles and how the approval process should be. Companies like Leica Geosystems AG set an authorisation process with a four-eye principle. This means that the manager of the employee as well as the process owner, in the case of SAP authorisations, are the ones who accept or reject a certain SAP access. The process *Manage Cybersecurity, Data Protection and Data Privacy* supports companies in terms of establishing the needed measures to be protected from cyber-attacks. This requires the development and implementation of a security platform that guarantees the privacy of a company's data. To achieve these objectives, a part of the technical solution, a company should also ensure i.e., that computers have bit-blockers installed and that employees are aware and trained to avoid potential attacks. The bit-blocker codifies the data existing in a personal computer, while the training part ensures the awareness of employees towards several cyberattacks, that are mostly sent through spam emails to employees. The analysis of an email by an employee, and its evaluation to see if this could be a thread before opening it, is crucial to avoid potential risks. The process *Manage International Trade, Tax & Legal* (see Figure 3.14) relates to the legal framework, corporate taxes, and indirect taxes. The legal department supports the organisation providing legal advice to foster the decision making in areas like operation, instruction, research, and administration (Schmitz, 2022, para. 1). For example, every time a company would like to sign a contract with a distributor, the legal department provides templates that can be used worldwide. In this way it can be guarantee that contracts standards are kept and observed. To this process belong also the corporate taxation of profits, and the appliance of indirect taxes when selling goods across countries. Apart from the compliance with taxes, companies need also to observe international trade regulations. IT, as part of shared services of corporate, defines, deploys, and manage the IT strategy and service offering. IT also fulfils and implements the needed support of the organisation for different applications like Windows, Android and IOS operation systems, SAP S/4 SAP HANA systems, Salesforce

CRM applications, etc. Without a proper functionality of these systems, the company will partly stay inoperative with subsequent costs implications. In this sense IT evaluates the risks of potential interruptions in these applications. The process *Manage Projects and Operations* contain the subprocesses Ideation Management, Portfolio and Program Management, Project Management, Corporate Operations Management, and Sustainable Operations. While the Ideation Management aims to analyse new ideas and requirements based on new concepts, the portfolio and project management focuses on creating a portfolio of projects related to those ideas. In this portfolio management projects are prioritised upon ideas of the employees. For example, ideas generated on how costs can be reduced in the company, could generate a portfolio of potential projects upon a prioritising is needed. This prioritisation could be based on the projects which are most cost effective in terms of savings. Project Management is the operationalisation of the projects starting from planning the projects, it continues with executing and monitoring of performance, and it ends up with invoicing and closing them. The management of corporate operations normally in big companies need an organisation called shared services. Shared services, where IT, HR, etc. are allocated, serve the entire divisions or business units in order to operate and orchestrate the business. For example, IT provides the technological platforms upon online business solutions can be built-up for the whole company. Also, sustainable operations should guarantee that the created platforms are valid for mid-term and long-term solutions upon the observation and fulfilment of environmental, health and security compliances.

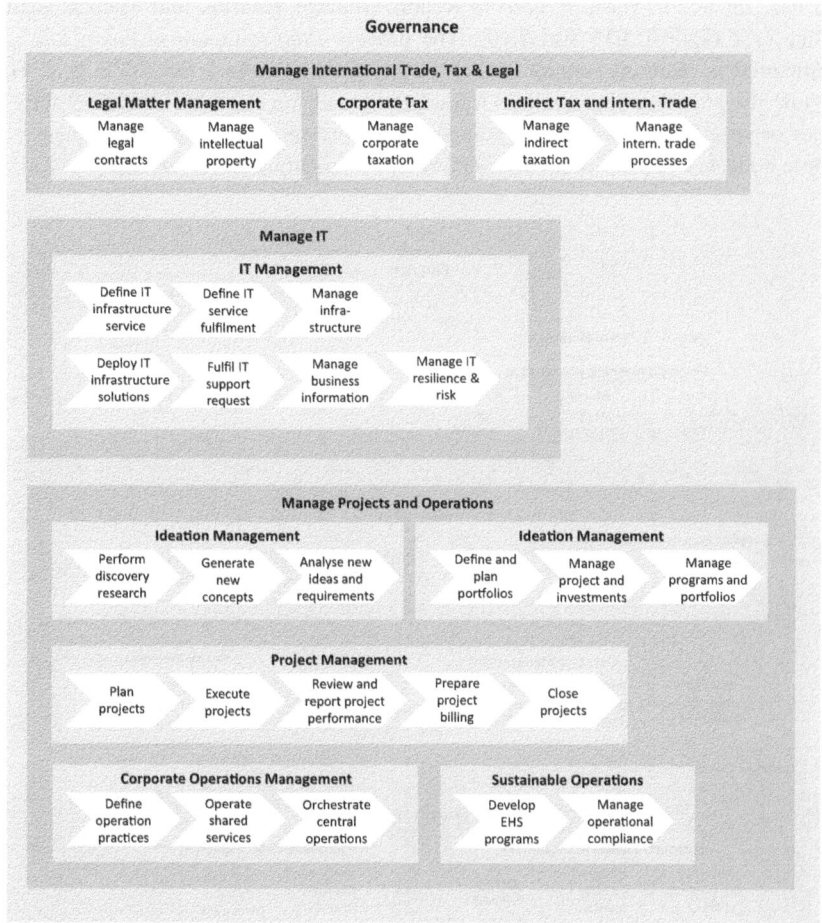

Figure 3.14   SAP S/4 HANA Governance process (Sarferaz, 2022, p. 217)

3.5.8   Process of Finance

The different business processes of a company interact with the process of finance, e.g., when selling goods and services, creating a purchasing order for the maintenance of an asset, repairing instruments for customers, renting an equipment, etc. Finance includes the processes of Plan to Optimise Financials, Invoice to Pay, Invoice to Cash, Record to Report, Manage Treasure and Manage Real State (see Figures 3.15 and 3.16). The process *Plan to-Optimise Financials* is structured in planning, operational and analysis phases (Sarferaz, 2022, p. 234). While the planning phase relates to budgeting and forecasting of business activities or specific period, the operational phase corresponds to the management of these budgets in the different divisions and business units (Sarferaz, 2022, p. 234).

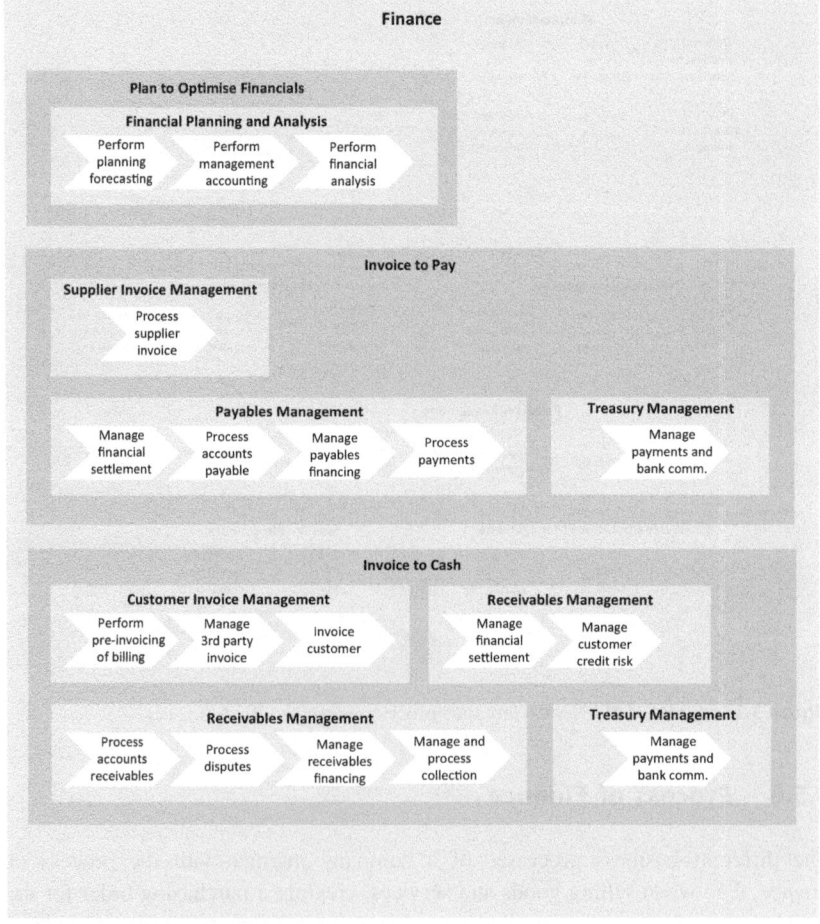

Figure 3.15 SAP S/4 HANA Finance process (Sarferaz, 2022, p. 234)

The analysis phase helps managers to monitor their budgets and see if they are going in a direction towards underachievement or overachievement. For example, reports related to billings and sales help sales managers to see if they are going to reach the budget for the quarter, as it was agreed with the top management of a company. The process *Invoice to Pay* covers the subprocesses related to the management of payables to suppliers and its operational realisation. In this process the finance department normally makes an invoice verification, if appropriate approves the invoice, posts the invoice, and finally does the payment (Buechler, 2018, para. 1). The process *Invoice to Cash*, on the opposite, refers to the management of receivables in terms of revenues and its daily operationalisation. The revenues generated from customer billings can either be recognised ad-hoc or monthly. In the case of a contract with customers, where they receive support and maintenance for an equipment for one year, the revenues need to be recognised for 12 months. The reason for this is that such a customer contract means a promise that the company needs to fulfil for one year. The process *Record to Report* comprehends the operational activities related to the creation of financial records, performance of financial accounting, financial closing and reporting (see Figure 3.16).

The main objective of the monthly financial closing is to report a consolidated financial statement to the stakeholders like e.g., the top management team and the controllers, as well as provide accurate management reporting (Buechler, 2022, para. 1). The reporting normally also includes the intercompany invoices. Intercompany invoices are the ones between the factory or headquarters and the selling units, or among selling units of an organisation. Thus, e.g., selling units can buy hardware and software from factories, selling units can repair instruments for other selling units, selling units can buy spare parts from distribution centers, etc. The *Manage Treasury* process supports an organisation by managing the payments, the communications, the cash and working capital, as well as securing financial risks and implementing treasury procedures and policies. Finally, the process *Manage Real State* helps the company to manage the real state of a company, starting with the definition and planning of a real state strategy. The subprocesses consider not the onboarding, but also the retirement of real state. It supports also the operationalisation related to the activities of the real state as well as assets.

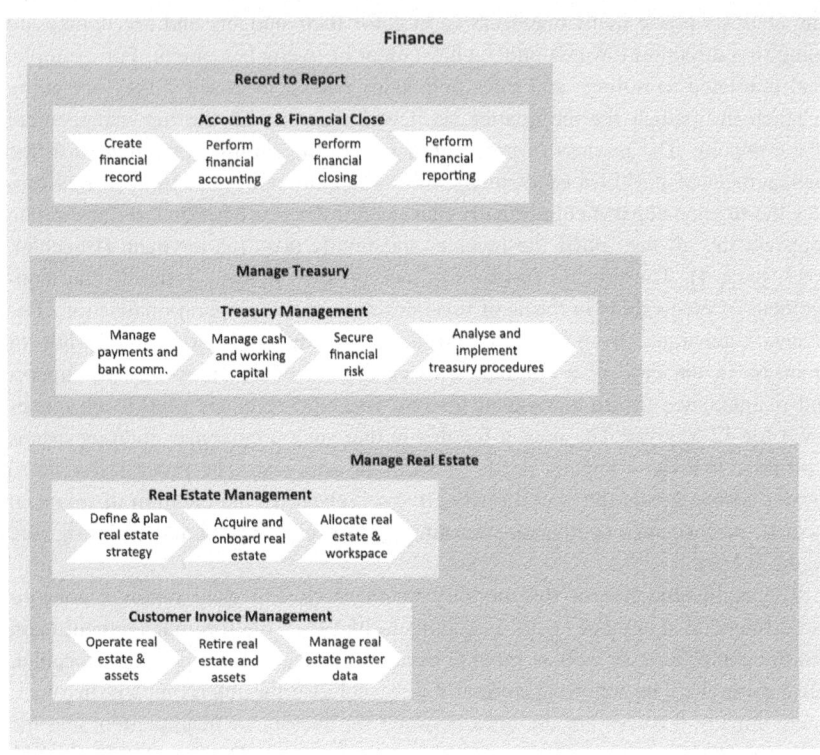

Figure 3.16 SAP S/4 HANA Finance process (Sarferaz, 2022, p. 235)

Knowledge Base

4

Prior to any design cycle of an e-learning artefact, the scientific knowledge base needs to be worked out to identify the theories and methods that will be the basement for the development and construction of the e-learning artefact, as well as the contribution to the DSR scientific knowledge. The knowledge base for this research project refers to the didactic theories and models, to the Information Systems Success Model (ISSM) requirements to build e-learning artefacts, and to the 3–2–1 expositional design model. These theories and models will provide the knowledge base grounding for the technical development of an e-learning artefact in the context of teaching and learning service software business processes for the new SAP S/4 HANA technology.

4.1 Learning and Didactic Theories

4.1.1 Learning and Didactic Theories. A Frame Comparison

This chapter aims primarily to give an overview about the differences between didactic and learning theories. Learning is a key concept in the educational sciences. Educational sciences could not be understood without learning (Reinhold, Pollak & Heim, 1999, p. 351), in the sense that educational activities take place through learning. However, the learning perspective in educational sciences and in psychology is different. While for educational sciences the learning situation of the learning person is key, for psychology the learning person is the focus of the learning (University Duisburg-Essen, Einführung in die Pädagogik, para. 16). Learning theories have their origin in the learning psychology while didactic theories have their origin in the educational sciences (Kron, 2000, p. 47) (Figure 4.1).

© The Author(s), under exclusive license to Springer Fachmedien Wiesbaden GmbH, part of Springer Nature 2023
F. Garayo Maiztegui, *Design and Evaluation of an E-Learning Artefact for the Implementation of SAP S/4 Hana®*, Gabler Theses,
https://doi.org/10.1007/978-3-658-40731-5_4

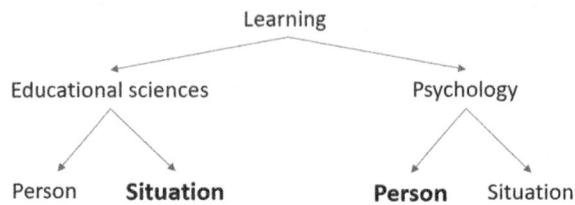

Figure 4.1   Different perspectives of learning theories and didactic theories (University Duisburg-Essen, Einführung in die Pädagogik, para. 16)

According to Kron (1999, p. 42) didactic is a science that focuses on the phenomena teaching and learning. In this area of didactic several didactic models have been developed. Didactic, as a science, also comprehends the legitimised, organised, and professional basis of teaching and learning processes (Kron, 1999, p. 44). Additionally, didactic theories explain the relationship among teaching objectives, its contents, and methods (Riedl, 2010, p. 11). On one side, didactic theories differentiate themselves in terms of how these teaching objectives, contents and methods are conceptionally structured, and how teachers with different roles act during the teaching process (Riedl, 2010, p. 11). Additionally, didactic models have also different scientific and historical origins. On the other side, learning theories try to identify mechanisms of the learning (Redaktionsteam PELe, 2006, p. 2).

There are basically three key learning theories: the behaviourism, the cognitivism, and the constructivism (Lehner, 2009, p. 96). These three different theories determine the way learning methods and concepts are defined. The learning theory of behaviourism explains that learning takes place thanks to an external influence, a so-called stimulus (Hubwieser, 2007, p. 3). This external influence produces a change in the behaviour of the learner. The core aspect of this theory is the relation between behaviour and stimulus (Hubwieser, 2007, p. 3). A learner changes certain behaviour or expected learning triggered by an external influence or stimulus. The external influence or stimulus can be positive as well as negative. While a positive stimulus would be e.g., the achievement of a grade, a negative stimulus could be represented by a reduction of monetary incentives in a company. For behaviourism the internal cognitive learning process of the individual is not relevant. The cognitive learning process, represented by the cognitivism, is contrary to the behaviourism. The cognitivism explains the behaviour of a human being based on a conceptual insight, this means that a learner can analyse problems and has the faculty of abstraction (Hubwieser,

2007, p. 3). This theory focuses on the learning as an internal process inside of a human being's brain (Lehner, 2009, p. 97). The brain of a human being processes and saves information continuously. A third theory is represented by the constructivism. The constructivism puts in the centre of this theory the ability of the own human being to perceive the environment and interpret it (Hubwieser, 2007, p. 10). Learning characteristics of the constructivism are (Hubwieser, 2007, p. 10):

* Learning is possible if a learner has the intrinsic will and interest of learning and actively takes part in the learning process.
* A learner takes responsibility and monitors his own learning.
* Learning depends on a learner's own know-how and experience, as well as the capability of interpreting information.
* Learning takes place on a specific context where learning can be then situational.
* Learning is a social process. On one side learners and teachers are subjected to a socio-cultural influence, while on the other side learning is an interactive process.

While the learning theories try to explain how the learner learns, the didactic theories try to get the learner to learn (Uljens, 2005, p. 37). Although both learning theories and didactic theories address the concept of learning, they do it from different perspectives and aspects. The following Table 4.1 shows a summary of some key differentiation aspects about learning and didactic theories.

Table 4.1   Aspects of the learning and didactic theories

Learning theories	Didactic theories
Its origin lays in the learning psychology	Its origin lays in the educational sciences
The focus is the person in the learning context (i.e., if the learner is getting bored or excited during the learning)	The focus is the happening of the learning situation and the factors that influence learning (i.e., it considers the happening in and around the whole class)
They try to identify the mechanisms of learning that takes place in the learners (i.e., if behaviour changes, faculty of abstraction, interpretations)	They explain the relationships among teaching objectives, teaching content and teaching methods
They explain how the learner learns	They get the learner to learn

4.1.2   General Didactic and Subject-Related Didactic

This chapter aims primarily to give an overview about the differences between general didactic and subject-related didactic. There has been an interrelation and on-going discussion between the general and subject-related didactic, especially since the 1970s (Plöger, 1999, p. 16). General didactic is the science of teaching and learning, and it comprehends the teaching and educational contents, as well as its structure, analysis, planning and implementation (Paape, Kiereta, & Maus, 2013, p. 36). Subject-related didactics (e.g., didactic implemented in business process engineering) try to combine the subject-related science with the general didactic theories (Plöger, 1999, p. 16). According to Peterßen (1983, p. 30) the subject-related didactic is part of the general didactic, and comprehends all efforts made in the areas of science disciplines to transfer subject-related knowledge like e.g., the usage of mathematics in engineering disciplines. According to Peterßen (1983, p. 30), the subject-related didactic is an abstraction of the general didactic and should orientate itself towards the science subject- related of the learning. The subject-related didactic tries to determine the learning objectives, the models of the content, the methods of the teaching, as well as to identify and prioritise new scientific didactic approaches for these learning objectives, models, and methods of teaching (Plöger, 1999, p. 17). In this sense general didactic and subject-related didactic should not be treated as something separate, but as a co-operation toward the common objective of educational teaching and learning processes (Kron, 2000, p. 33). Business processes as a subject in the area of business administration, as well as the learning and teaching of business processes, should be handled through a subject-related didactic perspective, where the combination of analysis, planning and implementation of the general didactic models interact with science themes of business administration in terms of business processes. When teaching business processes through an ERP System like SAP S/4 HANA, and utilising an e-learning platform, it is crucial to identify which general and subject-didactic theories should apply for the knowledge transfer of SAP business processes. In this sense section 4.2 below gives an overview of didactic theories and models, that at a later stage are the basis utilised in the development of the e-learning artefact. These didactic models should be considered as part of the knowledge base in DSR.

4.2   Didactic Theories and Models to Develop a E-learning Artefact for an Integrated Business Process Software

Didactic theories are declarative systems for structuring and planning teaching as well as learning processes (Riedl, 2010, p. 22). Didactic theories help to understand and explain the workflows, as well as relationships given in a teaching and learning process (Riedl, 2010, p. 22). They should enable a professional action-orientation approach based on a scientific basis, as well as provide three major functions: orientation, structure, and legitimation (Riedl, 2010, p. 22). Orientation means that the didactic theories should describe teachers' professional scope of duties and their corresponding challenges (Riedl, 2010, p. 22). It also means that teaching happens on a specific place, where specific content should be learnt (Riedl, 2010, p. 22). In this sense, teachers should choose objectives, content, themes, methods, and media to ensure that learning happens, and that learning results can progressively be monitored as well as evaluated (Riedl, 2010, p. 22). In a structural function, didactic theories help to make complex activities and challenges to those activities manageable (Riedl, 2010, p. 23). Thus, it provides a systematic platform to take decisions about the teaching procedure easier (Riedl, 2010, p. 23). Because every didactic activity causes a change on learners, the legitimation function focuses on legitimising specific teaching objectives, contents, and ways of teaching (Riedl, 2010, p. 23).

4.3   Didactic Theories and Didactic Models

Kron (2000) names several didactic theories and models that have been developed in the last 70 years classified in terms of education, learning and interaction. A didactic theory is by itself the final form of scientific knowledge (Martial, 2002, p. 123) and focuses not only on instruction or processes of teaching and learning in a specific sense, but it comprehends several aspects that influences the design of the teaching like e.g., condition of teaching, teaching plan, instructional plan, preparation plan, and pre- and post-analysis of teaching (Martial, 2002, p 123). A didactic theory building can contain more or several didactic models (Riedl, 2010, p. 72). According to Peterßen a didactic theory is fulfilled when it is developed on three structure levels (Lehner, 2009, p. 41):

- Paradigmatic level. This means a thinking approach based on a recognised scientific position, where the didactic field is researched.

- Legitimisation level. Reasoning for the pedagogical action in the sense of enhancing competences, autonomy, and solidarity.
- Pragmatic level. This means categories upon teachers can take decisions.

Teaching itself is a far too complex process to reduce it to a single didactic model (Martial, 2002, p. 124). A didactic model is a simplified or partial representation of teaching, that engages and limits itself to different aspects of teaching. Aspects of teaching could be for example, what is happening during the teaching in terms of structure, how is the communication, what learning process should be considered, etc. (Martial, 2002, p 123). The instruction of a lesson cannot be covered only from the perspective of a single didactic model, but it requires more the combination of several models that are supported on different theories (Martial, 2002, p. 115). The didactic of the learning theory, for example, combines both the didactic Berliner and Hamburger models, the cybernetic-communicative didactic model covers partially normative theoretical parts like learning objectives and principles (Martial, 2002, p 123). The following chapters describe several didactic models.

4.3.1 Didactic Models of the Educational Theories, Learning Theories, Constructive Theories, and Other Relevant Didactic Approaches

This chapter shows different didactic models that have been developed in the last 70 years mainly in German speaking countries. The Anglo-American literature addresses teaching and learning challenges normally without having a theoretical model building (Uljens, 2005, p. 33). The area of didactic is mainly larger than educational psychology and it includes much philosophical and theoretical thinking. In German literature, didactic and educational psychology are clearly separate fields with different representatives. The situation in Anglo-American literature from countries like Great Britain and the United States of America is quite the contrary; the same scientists are researching in this common area.

Thus, in the Anglo-American literature the theory of teaching is less developed because the empirical research aims not always to contribute to a conceptual system, and if it does, then the research in teaching comes closer to the German version of descriptive didactics (Uljens, 2005, p. 33). The Figure 4.2 shows an overview of the development from different didactic models in the last 70 years in German speaking countries. The next chapter will consider some key theoretical

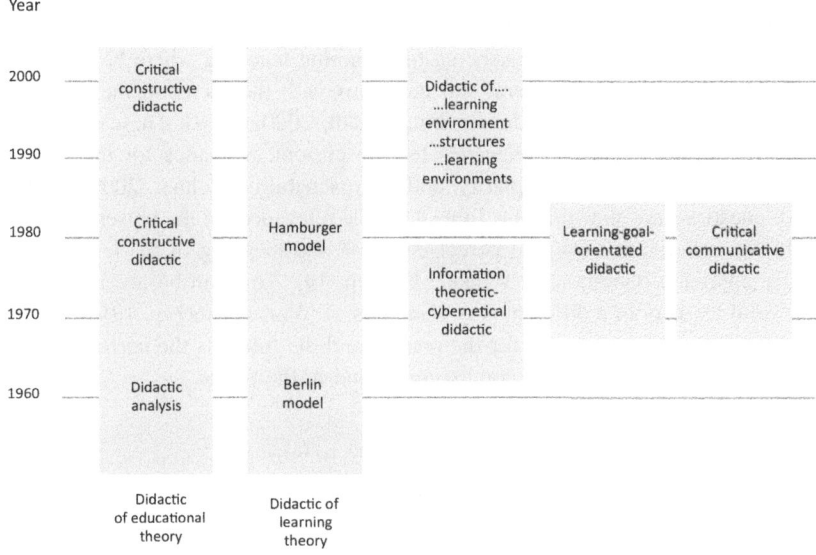

Figure 4.2 Development of the general didactic (adapted from Riedl, 2010, p. 83)

didactic models, based on their didactic theories, that are relevant when applying e-learning concepts. These learning concepts are the didactic analysis and
critical-constructive didactic model of the educational theory, the didactic models of the learning and teaching theory, and the constructive didactic models with
the cybernetic and information didactic, the learning-goal-orientated teaching, the
critical and communicative didactic and the action-orientated didactic.

4.3.1.1   Didactic of the Educational Theory
4.3.1.1.1   Didactic Model of the Didactic Analysis
Wolgang Klafki developed the didactic analysis as a model of educational theory
in the 50s (Lehner, 2009, p. 74). The didactic model of the educational theory
focuses on a general educational approach, and on the choice of the teaching
content (Lehner, 2009, p. 74). The whole didactic analysis is based on a ped-
agogic of humanities with the primary aim of education (Riedl, 2010, p. 88).
Education in the didactic analysis is understood how the learner encounters the
cultural environment, being part of it, and at the same time adopting it (Riedl,
2010, p. 88). The didactic analysis tries to answer questions related to which

content is appropriate for education and what is the best way to transfer this content through a teaching planning in a class (Riedl, 2010, p. 89). In 1985 Klafki published a perspective schema for planning teaching where he included five major questions with several sub questions with the focus on the teaching content as well as subject of the teaching (Riedl, 2020, p. 89). These questions refer to several aspects like relevance for the present, relevance for the future, structure of the content, exemplarity, and approachability (Lehner, 2009, p. 76). These questions are shown in the Table 4.2. The relevance for the present and for the future is determined by the perceived sense and meaning of the learners on a daily approach (Gudjons, & Winkel, 1999, p. 19). This can be ascertained in by several learners in a different way (Gudjons, & Winkel, 1999, p. 19). The difference between the relevance for the present and the future is the corresponding timeline in terms of the social context today and in the future.

Table 4.2 Didactic analysis questions (Lehner, 2009, p. 74)

Aspect	Question
Relevance for the present	What is the relevance of a specific theme for learners in the present to gain experience, awareness, competence, and skills?
Relevance for the future	What is potential relevance that a theme might have for the future?
Structure of the content	What is the structure of the content?
Exemplarity	Which context and factual meaning should the content have? (Which methods, techniques can be considered?)
Approachability	Which are the special situations, phenomena, persons, etc. that make the class for the learners interesting?

The structure of the content refers to the structure of the themes, which also include the learning methods (Gudjons, & Winkel, 1999, p. 25). Both methods and structure go together hand in hand with the purpose that learners acquire knowledge, e.g., solutions possibilities given for a specific mathematical problem (Gudjons, & Winkel, 1999, p. 25). The context and factual meaning of the exemplarity is related to the learning objectives, which can contain several sub learning objectives (Gudjons, & Winkel, 1999, p. 21). Based on these learning objectives, the description of frame methods are needed to achieve such learning objectives. A further question to be considered is the approachability. Approachability is about how certain knowledge can be approached by the learners in terms of representations, actions, simulations, games, etc. (Gudjons, & Winkel, 1999, p. 21).

Additionally, to these major questions, Klafki also defined how the educational content can be thematised between general and special aspects (Lehner, 2009, p. 75). The main objective is to recognise learning and experience possibilities of the learners with the following classification (Lehner, 2009, p. 75):

- Elementary knowledge, e.g., the basic principles of a physical law.
- Fundamental knowledge, e.g., basic knowledge transfer of basic experiences and insights in the environment like how a butterfly flies.
- Exemplary knowledge, e.g., knowledge transfer through the feeling experience of a specific time visiting the museum of Elvis Presley in Graceland.

4.3.1.1.2 Critical-constructive Didactic

In the 80s Klafki developed this model further to a critical and constructive didactic (Lehner, 2009, p. 75). "Critical" means that teaching preparation concepts should achieve three main objectives: self-determination, co-determination, and solidarity (Lehner, 2009, p. 76). While the self-determination relates to the personal social relationships and sense making among human beings, the co-determination refers to the personal responsibility in the social and political construction (Plöger, 1999, p. 73). Solidarity is understood as ethical corrections when the self-determination and the co-determination is not achieved, either because social relationships are missing, or underprivileges, political limitation or social oppression come ahead (Plöger, 1999, p. 73). "Constructive" refers to an open mind perception away from a rigid structure of teaching design concepts and towards the acceptance of changing learning environments in practical teaching (Lehner, 2009, p. 76). It is a process of reflexion where a constructive open minded is needed. With this approach the teachers can adapt themselves to unpredicted situations, where a rigid structure might not work. The idea of Klafki at the time of developing a new critical and constructive didactic was to provide teachers with a planning method that could help them to structure their teaching (Riedl, 2010, p. 95). The critical and constructive didactic of the educational theory aims to structure the teaching and the planning, and is based on a didactic analysis, but considers further major structural aspects like the analysis of the conditions, the methodical structure and the proof and monitoring of the learning success (see Figure 4.3).

According to Kron (2000, p. 136), the analysis of the conditions refers to the social cultural initial conditions of a group (e.g., a class), to the teaching relevant institutional conditions (either changeable or not changeable in a short term), and to the possible difficulties or disturbances in the environment. All these types of conditions, before the teaching and learning takes place, might influence positively or negatively in the learning of participants. The methodical structure is related to teaching and learning processes, as well as the needed materials or media aspects of the teaching (Kron, 2000, p. 136). This also represents the different interactions with different steps in the learning process structure (Gudjons, Winkel, 1999, p. 30). Finally, the proof and monitoring aims to identify if the learning of the content has taken place, as well as to confirm that the teaching and process method has its practical validity.

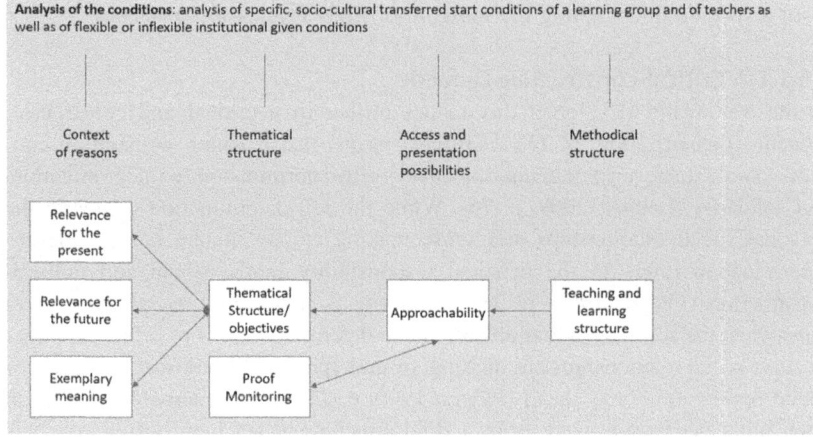

Figure 4.3   Perspective schema for the teaching planning (adapted from Klafki, 1999, p. 18)

4.3.2   Didactic Models of the Learning Theory

Paul Heinmann is seen in the literature as the founder of the learning theory (Lehner, 2009, p. 77). Heimann believes that the planning of the teaching should not be based on education, but on learning (Riedl, 2010, p. 101). He postulates that teaching should be more practical and criticises those teachers from the educational theory who act far away from the practical reality of teaching (Riedl, 2010, p. 101). In opposition to the educational theory, the learning theory is empirically oriented (Lehner, 2009, p. 78). Heimann criticises that the educational theory in a narrow perspective that only considers content aspects, and that education as a central aspect is inadequate for the planning of the teaching (Riedl, 2010. P 102). Heimann postulates that the learning theory should be based in three education reasons (Plöger, 1999, p. 105):

• in the teaching analysis for the common participation on classes,
• in the planning of the teaching purposes triggered by students,
• and in teaching experiments, with the aim of verifying or falsifying didactic hypotheses.

4.3.2.1 The Berliner Didactic Model

Because teaching, in its early stages, was considered to be far away from the practical world, Heimann, Gunter Otto und Wolfgang Schulz created the so-called "Berliner didactic model" (Lehner, 2009, p. 68). They believe that teachers need a learning scientific theory, upon the teaching is built-up and corrected, as well as the theory is developed (Stracka, & Macke, 2002, p. 25). The authors created a model for the planning and structural analysis of teaching. In this sense they developed a concept that follows different interests like the situation of the teachers, teaching under the cohesion of several conditions, the structural analysis, the factor analysis, the factors correlation, and the learning to initiate (Kron, 2000, p. 138). They believe that teachers not only should plan the teaching, but they also should in parallel analyse it (Kron, 2000, p. 138). In this sense the authors assume following implications (Kron, 2000, p. 138):

• teachers are at the same time researchers,
• teaching cannot be interpreted without the context or environment where it takes place,
• and planning, as well as teaching, do not happen without analysis.

The authors affirm that teachers should be coherent in terms that teaching needs adequate research, and research scientific methods like teaching experiments, observations, reflections, and experience should apply (Kron, 2000, p. 138). The research is not only limited to these methods, but it also should include the research of individual, collective and society relationships (Kron, 2000, p. 138). Another important aspect is the structural analysis of the teaching. The basic idea of the structural analysis is to understand, which factors do play a role in teaching and how these should be combined to ensure sustainable learning (Lehner, 2009, p. 78). These factors (see Figure 4.4) are classified in four decision and in two conditions fields.

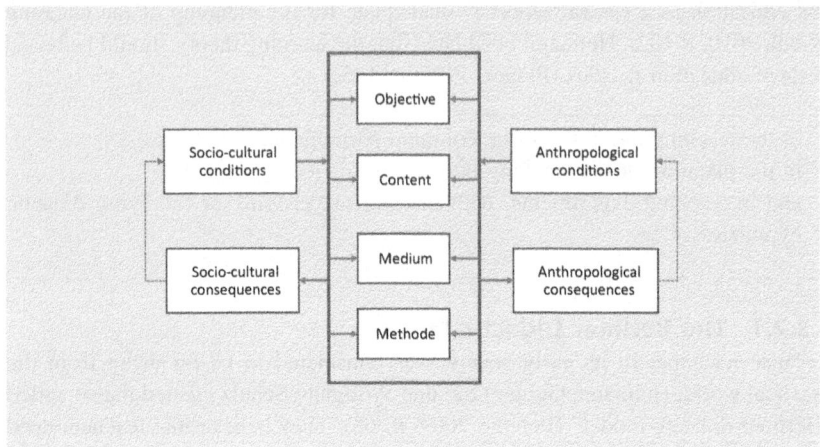

Figure 4.4 Factors of the Berliner model for the structure analysis of teaching (adapted from Lehner, 2009, p. 79)

   While the four decision factors are intentions, content, methods, and media, the two condition factors of the Berliner model refer to the anthropologic and the socio-cultural conditions. The so-called intention factor relates to the learning objectives and purposes formulated in every learning and teaching process to make these learning objectives transparent and explicit (Lehner, 2009, p. 78). The content refers to sciences, techniques, or certain ability to act (Riedl, 2010, p. 106). Additionally, to the factor content, also the methods of the learning and the teaching process should be a representation and an expression of the professionalism of teachers (Lehner, 2009, p. 78). The selection of these methods is conditioned by categories like articulations (phases of teaching), group and room organisation (relationship between teachers and learners), teaching and learning approach, different methodical models, and didactic principles (Riedl, 2010, p. 106). Finally, the factor media (e.g., documentation, sound, videos, etc.) also plays an important role for supporting the communication between teachers and learners (Lehner, 2009, p. 78). The anthropological and socio-cultural conditions are the beginning conditions of the teaching (Riedl, 2010, p. 106). The condition factors related to the anthropological conditions are defined e.g., by the learning, language conditions, learning style, etc., while the condition related to the socio-cultural aspects are determined by the size of the group, the cultural background, the culture in communicating, etc. (Lehner, 2009, p. 78). The factor analysis looks for the conditions that factually influence the teaching (Kron, 2000, p. 140). These conditions are related to society in the sense of norms and rules, to conditioning factors like sciences and education, and to organisational factors like traditional teaching methods and models (Kron, 2000, p. 140). The authors define the correlation within these factors as an open system (Kron, 2000, p. 140). This open system should be able to set conditions of the teaching that the teachers need to know to prepare and plan the teaching, to conduct it, and to evaluate it (Kron, 2000, p. 140). All didactic endeavours should serve the initiation of learning with the election of a learning theory, where the learning is not seen only as a fix compound of structural elements, but as a process (Kron, 2000, p. 140).

4.3.2.2 The Hamburger Didactic Model

A further development of the Berliner didactic model represents the Hamburger didactic model developed by Wolfgang Schulz, a disciple of Paul Heinmann (Stracka & Macke, 2002, p. 33). The Hamburger didactic model sees all parties (teachers and learners) involved in the teaching as active stakeholders for the planning of a class lesson (Riedl, 2010, p. 110). The Hamburger didactic model can be described in five steps: the criteria for the planning, the structure moments of the didactic actions, functions of the didactic, the principal thoughts of the planning, and the levels of the planning (Kron, 2000, p. 145). The criteria for the planning are defined through life orientation, science and subject didactic orientation, actions orientation, methods orientation, orientation towards the individual and collective hindrances of the learning processes, orientation towards the media and the organisation forms of teaching, and orientation towards self- and external control (Kron, 2000, p. 145). The structure moments of the didactic actions are determined by the criteria of the planning: understanding among teachers and learners, the definition of the teaching objectives and the initial situation, the determination of the knowledge transfer and success variables, the working out of the institutional conditions and the perception of the social contradictions (Kron, 2000, p. 145). The structure moments of the didactic actions are represented in Figure 4.5 through political and economic relations, the class related actions of own and world understanding, and the institutional conditions. Schulz defines the functions of the didactic as actions that teachers should approach to become professional in what they do (Kron, 2000, p. 147). These are defined through actions in the sense of advice, assess, analyse, plan, execute, administer, and be co-operative (Kron, 2000, p. 147). The principal thoughts of the planning are conditioned through the thematical, personal and social association aspects

(Kron, 2000, p. 147). While the thematical aspects are determined by the intention to consider social themes with qualification purposes, the personal aspect is represented through the intention of individuals to work themselves out through the different themes (Kron, 2000, p. 147). The association aspect is identified through the intention of group members to interact with each other and benefit from each other's support, as well as thematical and personal development (Kron, 2000, p. 147). Schulz distinguishes four levels for the didactic planning: the perspective planning, the outline planning, the process planning and the planning correction (Riedl, 2010, 112). The perspective planning should be a help for the teachers, based on a dialogue with others, to define the content aspects and prioritise them, look for methodical and media important influence factors, and thus define specifically the different class lesson units (Riedl, 2010, p. 113). The outline planning, seen as many in the scientific world as the core of the Hamburger model, tries to define which teaching objectives are based on which themes, and with which learning procedures (Riedl, 2010, p. 113). The Figure 4.5 shows in the core of the outline planning, that the training takes place among teachers (T) and learners (L). The representation of more than one teacher and learner means that teachers plan different trainings with several groups of learners (Riedl, 2010, p. 113). The brackets between teachers and learners represent in Figure 4.5 that both teachers and learners should have a commitment in all dimensions of the teaching to ensure that learners get involved as active contributors of the class lesson (Riedl, 2010, p. 114). The actions of teachers and learners depend on the sociocultural conditions in terms of institutional conditions (Martial, 2002, p. 166). Between the brackets following dimensions for the learning and teaching are considered: objectives (OB), initial situation (IS), success monitoring (SM) and transfer variables (TV). While OB mean objectives in terms of intentions and themes of the teaching, IS refer to the initial situation of the teachers and learners (Riedl, 2010, p. 115). SM approaches the self-monitoring of teachers and learners to define the influence of the teaching, while TV refer to the methods and media of the teaching (Riedl, 2010, p. 115). The actions taken by teachers and learners in the context of learning will depend on the sociocultural conditions represented in the Figure 4.5 as institutional conditions (Martial, 2002, p. 166). These institutional conditions are determined at the same time by the political and economic context, as well as how they are understood (Martial, 2002, p. 166). The process planning defines several different teaching steps, methods, communication and working forms (Riedl, 2010, p. 116). In the process planning it is crucial to describe aspects like needed time, a sequential and chronological teaching plan, as well as learning objectives and learning monitoring mechanisms (Riedl, 2010, p. 116).

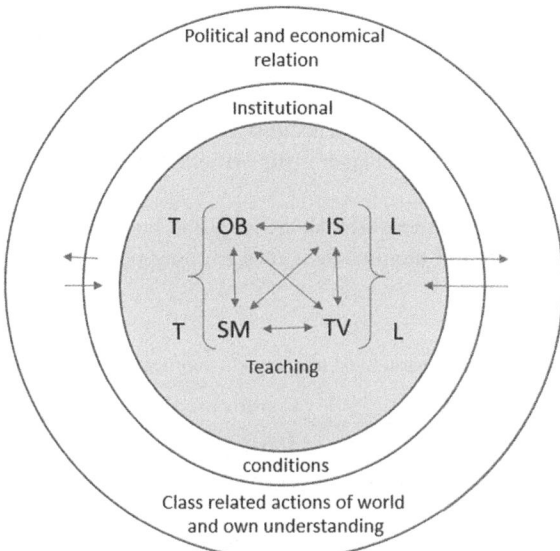

Figure 4.5 The outline planning (adapted from Riedl, 2010, p. 114)

The planning correction should be the result of teaching reflections made by the teachers after the teaching (Riedl, 2010, p. 116). In this sense teachers can make corrections, modifications, or improvements for the planning of the next class lesson. The teaching becomes a continuous improvement process.

4.3.3 Constructivist Didactic

The constructivist didactic addresses the organisation of teaching and learning processes where all possible thinking and action are orientated towards the learners´ active and autonomous construction of knowledge (Riedl, 2010, p. 118). In comparison to other didactic theories, the constructivist didactic focuses on the individual, where teachers organise the learning environment and take the role of a coach for individual learning processes (Riedl, 2010, p. 118). The learning environment is decisive for the definition of the teaching methods and techniques, learning materials, media (Riedl, 2010, p. 119). The constructivists learning environment differs in many aspects from an objectivist learning environment (see Table 4.3).

Reinmann and Mandl (2001, p. 626) argue that learning should be active, self-steered, situational, and social. The authors postulate that the learning should contain the following process characteristics (Riedl, 2010, p. 125):

• learners are active and motivated about what and how they do it,
• learners must steer and monitor their own learning processes,

Table 4.3 Objectivist and constructivist learning environments (Riedl, 2010, p. 119)

Objectivist view	Constructivist view
The focus is teaching.	The focus is learning
Learning is the result of the teaching organisation from teacher's view.	Learning can only take place through the learners.
A good teaching process triggers a good learning process.	Teaching is understood as a support of the individual self-learning.
Learning is a linear, systematic, and deterministic process, where teachers present and transfer clear and structural learning contents.	Learning is a non-linear, multi-dimensional and non-deterministic process, where learners integrate complex and dynamic learning content in their own knowledge and actions structures.
Teachers monitor the learning results of the learners.	Teachers organise, mentor and coaches learning environments.

• learning takes place in a constructive way, being aware about the existing experiences and knowledge of the learners,
• learning is situational, depending on specific learning environments,
• and learning is a social process, where the social interactions and socio-cultural backgrounds are considered.

There are different didactic constructivists approaches like the anchored instruction approach, the cognitive flexibility theory, and the cognitive apprenticeship approach. The anchored instruction has as a priority the avoidance of passive knowledge (Reinmann, & Mandl, 2006, p. 629). The key concept lays in a so-called "narrative anchor", where learners identify authentic problem situations and are interested to solve them (Reinmann, & Mandl, 2006, p. 629). The layout principles are (Reinmann, & Mandl, 2006, p. 629) characterised by the presentation of authentic problem situations visualised in the form of videos to get the interest of the learners, by a problem represented in an environment or context familiar to the learner, by all data needed to solve the problem included in the video, by the complexity of the problem related to a real situation, by the problem shown to the learner in two different versions with the aim of showing different perspectives for a flexible problem solving. The cognitive flexibility theory is an approach related to the expertise research, where the focus is to make the learners familiar with complex real situations and avoid a simplistic approach of the issue (Reinmann, & Mandl, 2006, p. 630). The main objective of the cognitive flexibility approach is that learners learn in multiple directions and perspectives, and that the knowledge is applicable in different facets and in a flexible way (Reinmann, & Mandl, 2006, p. 630). In the cognitive apprenticeship approach learners will be integrated in a problem-solving environment, where they start from the very beginning to learn what it is needed to solve a specific problem. In this kind of situation, the problems become always more complex, and the learners learn to apply their knowledge in more flexible ways. The way learners co-operate and work together with teachers and another learner enables them to become part of the expertise culture.

4.3.4 Other Relevant Didactic Models

4.3.4.1 Didactic of the Learning-Goal-orientated Teaching

The learning-goal-orientated teaching, also known as curricula didactic, was develop by Christina Möller (Riedl, 2010, p. 86). This model aims to monitor learning and teaching procedures, as well as avoid uncontrolled teaching related decisions (Riedl, 2010, p. 86). The model focus on the learning process (Korn, 2000, p. 157) and sets as priority the precise definition of objectives embedded in a hierarchical structure (Riedl, 2010, p. 86). The curricula didactic sees the learning process as a change on behaviour, that is produced through the sequence of learning phases (see Figure 4.6). The planning of teaching planning in this model contains three steps: the learning planning, the learning organisation (learning phases) and the learning monitoring (Kron, 2000, p. 160). The learning planning represents the "to-be" values that should be achieved (Kron, 2000, p. 158).

The step of the learning planning especially considers the definition of the learning objectives in terms of objectives for the cognitive learning, for the affective learning, and for the psychomotor learning (Kron, 2000, p. 159). For the

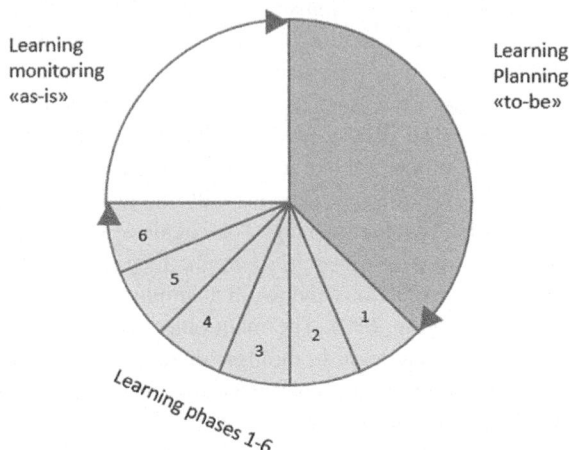

Figure 4.6 Curricula development as a control loop (adapted from Riedl, 2010, p. 86)

definition of these three types of learning the authors utilise existing templates from previous scientific publications like the cognitive learning objectives from Bloom, the affective learning objectives from the behavioural psychology from Pate, or the psychomotor objectives from Guilford (Kron, 2000, p. 159). Möller defines four steps at the time of defining the objectives: collection, definition, arrange and decision of the objectives for the planning of the teaching. The learning organisation or learning phases is based on the description and arrangement of the learning methods and selected media (Kron, 2000, p. 161). The definition of the learning organisation sets the objectives of the learning for the different phases as well as the decision about the media and learning methods based on defined objectives (Riedl, 2010, p. 87). The decision of the methods and media takes place during the definition of the objectives (Riedl, 2010, p. 87). The learning phases are the steps to achieve the learning objectives (Kron, 2000, p. 160). The learning monitoring aims to compare the "to-be" with the "as-is" learning objectives (Kron, 2010, p. 161). The development of the learner's behaviour is described through the statuses before ("as-is") and after the learning ("to-be"). The learning process takes place during the learning phases.

4.3.4.2   Critical and Communicative Didactic
The critical communicative didactic was developed by Winkel and is related to two communication dimensions (Kron, 2000, p. 189). First, it relates to a process

dimension based on different communications axioms like i.e., the communica-tions axions of Watzlawick (Kron, 2010, p. 191). This process dimension refers to the content and the relationship aspects of the communication (Kron, 2010, p. 191). In comparison to Watzlawick, Winkel gives the relationship and the content aspect of the communication the same value in the sense that learners should have a participation in the planning of the teaching (Kron, 191). Second, it relates to an enhanced communication dimension in terms that teaching and learn-ing communication must be transparent, co-operative and with few interferences (Winkel, 1999, p. 95). Winkel added the adjective "critical" because he believes that educational institutions should not take their current socio-cultural values for granted, but bring them under the criticism umbrella towards an improvement (Winkel, 1999, p. 94).

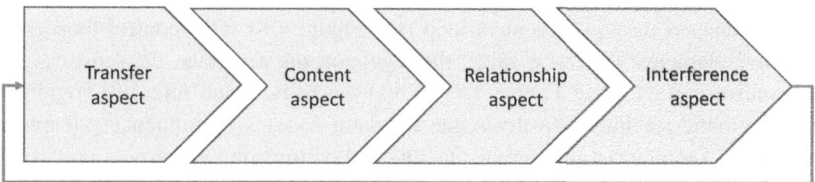

Figure 4.7 Teaching in the context of the critical communicative didactic (Winkel, 1999, p. 101)

Winkel developed four structural processes for teaching (see Figure 4.7) that can be of help for the analysis and planning of teaching (Winkel, 1999, p. 94). These are the transfer, content, relationship, and interference aspects. Under the transfer aspects in the following sub-aspects are considered: (1) learning defi-nitions and acts, (2) media, teaching and exercises, (3) teaching methods like experiments, individual work, teamwork, and peer group, (4) teaching structure and (5) teaching organisation (Kron, 2010, p. 192). The content aspect shows what is presented during the teaching, while the relationship aspect considers ele-ments of social interaction (Kron, 2010, p. 192). Finally, the interference aspect is related to different types of interferences like e.g., discipline issues, provocation, acoustic and visual interferences, learning refusal, and passive behaviour.

4.3.4.3 Cybernetic and Information Didactic

The cybernetic and information didactic was developed by Felix von Cube (Von Cube, 1999, p. 57). The main objective of this model is to increase efficiency in the didactic processes and thus consequently optimise educational as well as learning processes (Riedl, 2010, p. 83). The model tries to represent a teaching reality that describes teaching processes with a mathematical model (Kron, 2000, p. 149). The cybernetic theory finds its background in Skinner's behaviourism as learning theory, where monitoring elements should cause changes in the behaviour of individuals (Kron, 2000, p. 149). In this sense learning is understood as a change in the behaviour (Kron, 2000, p. 149). However, von Cube also utilises for his model the information theories from Wiener with the sender and receiver models (Kron, 2000, p. 150). In this sense cybernetic tries to answer the questions related to how information is transferred and monitored, as well as which type of information or signals are available (Kron, 2000, p. 150). Thus, the cybernetic and information didactic is understood as an individual and social system, as well as coherences of reality, whose elements are in interaction with other elements to achieve an equilibrium in the form of a regulator system (Kron, 2020, p. 150). This model sees education as a regulator system (Riedl, 2010, p. 84) managed through a control loop (see Figure 4.8). This control loop contains five elements: the to-be value, the regulator, the actuators, the sensors, and the control variable (see Figure 4.8). While the to-be value refers to cognitive and pragmatic teaching objectives, the regulator focuses on influencing learners in terms of keeping out interference factors and optimising the learning processes (Kron, 2000, p. 152). In this focus the regulator follows certain learning strategies to achieve the learning objectives (Riedl, 2010, p. 84). These strategies have big influence in the learning process, if they are less time consuming, and if they bring security in achieving the learning objectives (Riedl, 2010, p. 84). The actuators mean the personal and technical media to control the teaching process (Riedl, 2010, p. 84). The control variable is the learner (Riedl, 2010, p. 84). The learning improvements of the learners will be monitored by the sensors elements and its results compared in terms of to-be and as-is values (Riedl. 2010, p. 84).

According to Von Cube the cybernetic and information didactic can be implemented in the following areas (Kron, 2020, p. 152): learning can be determined as information processing, the cybernetic model can optimise the learning strategies, and learning programs with different media types can be implemented. The cybernetic and information didactic plays nowadays a secondary role due to a reductionism approach in currently complex teaching situations (Riedl, 2010, p. 84).

4.3.4.4   Action-orientated Teaching
The action-orientated didactic is defined as a system to straighten out and improve didactic praxis (Kron, 2000, p. 186). A learning situation is defined as the smallest research unit in phenomena-logical studies and empirical research as well as

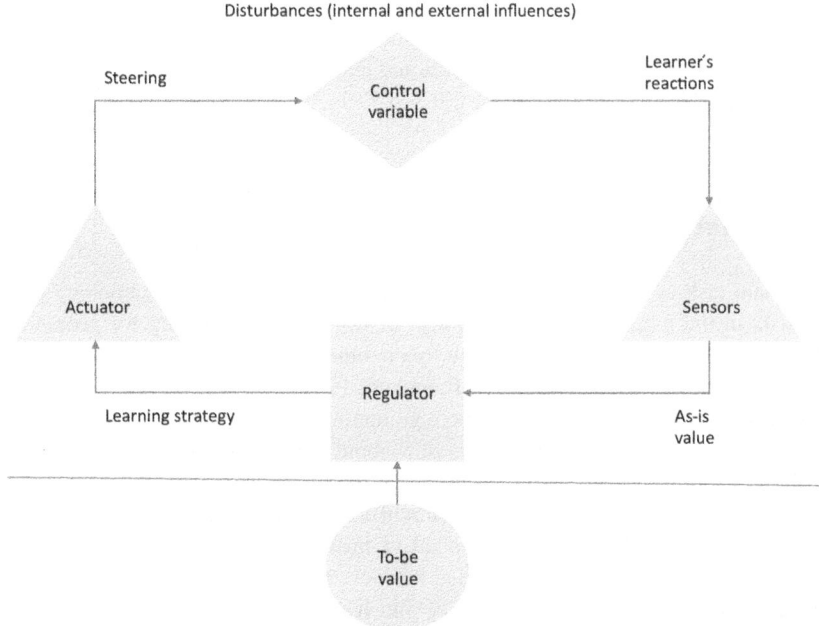

Figure 4.8   Teaching as a control loop (adapted from von Cube, 1999, p. 60)

a phenomenon, where several people are in an action-orientation relationship that goes beyond the classical definition of a scholar class (Kron, 2000, p. 186). The action-orientated didactic in the praxis applies also in other pedagogical and didactical critical learning areas like e.g., adult teaching and companies' development centres (Kron, 2000, 186). The action-oriented didactic requires the implementation of open models, where technical and practical actions can be applied (Kron, 2000, p. 187). This implies that teachers have a clear knowledge about learning theories and about learning taxonomies like learning by doing, learning by game simulation, learning by research, learning by teaching, learning by audio-visual media, etc. (Kron, 2000, p. 187). The action-orientated teaching focuses primarily on the transfer and promotion of key qualifications that should help learners to prepare themselves professionally, to take responsibility, and to think about the consequences of their actions (Paape, Kiereta, & Maus, 2013, pp. 65–66). Specifically, professional activities and roles throughout commercial activities in business processes require process orientated thinking, as well as capabilities to be able to identify if problems are supposed to be solved, and if the knowledge is supposed to be continuously built-up. In this context learners are active through the learning processes and teachers take the role of a consultant or coach (Paape, Kiereta, & Maus, 2013, p. 69).

4.3.5 Subject-didactic for Business Processes in Information Technology

The main task of the subject-didactic in the area of information technology (IT) consists in the research and development of learning and teaching for people of all ages (Baumann, 1996, p. 45). Its main objective is the design and further improvement of the teaching in IT (Baumann, 1996, p. 45). IT contains however a wide spectrum of different subjects (e.g., mathematical logic, date management systems, etc.). Table 4.4 gives an overview about the different fields of IT as well as their description. Applied IT business informatics is defined as the integration of business data and workflows in different areas like procurement, production, sales, etc. (Eberle, 1996, p. 38). Applied IT business informatics like Enterprise Resource Planning (ERP) as integrated business software is a solution to support the digitalisation of business processes and it covers different business process areas in a company like e.g., purchasing, production, sales, development, assets management, human resources, finance and controlling, quality management and project management (Körsgen, 2001, p. 161–162). Especially in the IT business informatics area, Rohs and Mattauch (2001, p. 5) developed a new work process orientated model called APO-IT ("Arbeitsprozessorientierte Weiterbildung in der IT" or Work Process oriented further training in the IT) that is defined through three major characteristics: orientation towards specific work processes in certain roles in a company, integration of work and learning, and increase of self-organisation of the learners.

While the orientation towards specific work processes extracts the importance of IT business processes related competences, the integration of work and learning aims to focus on the professional development of employees based on specific and individual learning objectives and learning content (Rohs and Mattauch, 2001, p. 61). The increase of self-organisation of the learners represents the digitalisation of the learning processes, where learners can access on demand multimedia material embedded in an e-learning environment, and with the coach co-operation of the trainer (Rohs and Mattauch, 2001, p. 62). APO-IT addresses also essential questions in terms of methodical and didactical thoughts like e.g., how will learning objectives and learning content be defined and justified, as well

Table 4.4 Overview about the fields in Information technology (Eberle, 1996, p. 36)

Information Technology (IT)	Description
Core IT	
Theoretical IT	Research and learning objects of mathematics including mathematical logic
Practical IT	Software orientated subjects for the architecture, programming, and data management systems
Technical IT	Relevant hardware concepts and theories of physics as well as electronics including digital memories and arithmetic units
Applied IT	
Application of IT	Adjacent subjects where IT applies, e.g., business administration, business informatics, economics, engineering, education, law, etc.
Social references IT	Questions about the social requirements and influences because of applying IT.
IT didactics	General didactics for different fields in IT.

as how should theory and praxis be combined (Rohs and Mattauch, 2001, p. 63). Process orientation and development of professional competences are the core of a modern and future orientated professional training (Otte, & Schmidt, 2008, p. 23). This development plays an important role, because business processes are no subjected to a defined bundle of sequential activities that are kept unchanged forever. On the opposite, business processes keep continuously changing as customers' requirements also do (Otte, & Schmidt, 2008, p. 25). Therefore, the subject-didactic of business processes contains a method where the teaching of business processes and SAP ERP systems are integrated, since business processes nowadays are processed and mostly taught through the usage with ERP systems (Otte, & Schmidt, 2008, p. 25). Also, the teaching and learning of integrated business processes with the software SAP S/4 HANA goes beyond clicking onto or maintaining data fields in a software platform. It requires primarily knowledge of business processes and workflows, as well as knowledge of technical information systems. Without this, an effective learning could not take place (Körsgen, 2001, p. 10). The subject APO-IT didactic, combined with the general didactic, will be part of the didactic method described in section 5.2.

4.4   Requirements for an E-learning Artefact

As part of the DSR knowledge base, the requirements from scientific methods for creating an e-learning artefact are also essential for this research project. In the academic literature there are several models, based on scientific results, that try to identify requirements for construction of an E-learning platform. These requirements also measure the success factors when implementing E-learning platforms. The following models have been identified in the academic literature: the Information System Success Model (ISSM) from DeLone & McLean (1992, p. 60), the Technology Acceptance Model (TAM) from Davis (1989, p. 319), the Flow Theory Perspective from Choi, Kim and Kim (2006, p. 223), the users' satisfaction models like e.g. Blended E-learning System (BELS) from Wu, Tennyson and Hsia, (2010, p. 155), or the model of Sun, Tsai, Finger, Chen, Yeh (2008, p. 1183), or the model from Paechter, Maier and Macher (2009, p. 222), as well as the quality focused model from Lee and Lee (2008, p. 32). However only some or extensions of these models have been used to measure success factors of e-learning with ERP systems. These models are the ISSM Model from DeLone and McLean (Paa, 2014, p. 125) and the flow theory perspective from Choi, Kim and Kim (2006, p. 224). DeLone & McLean provide with their Information System Success Model (ISSM) a model with six dimensions, that represent a framework for an e-learning concept in terms of critical success factors (Paa, 2014, p. 40).

They reviewed 180 scientific papers and examined 90 empirical studies found in academic literature between 1992–2007 (Stacie, DeLone, McLean, et. al, 2003, p. 236), and since then it has been in use in 285 scientific papers (DeLone, & McLean, 2003, p. 11). The six dimensions of the ISSM model are: system quality, information quality, service quality, intention to use or use, use satisfaction, and net benefits (DeLone, & McLean, 2003, p. 249).

These dimensions are interdependent to one another (DeLone, McLean, 2008, p. 238). Table 4.5 shows the different dimensions that influence a successful E-learning artefact. Based on these six dimensions, Paa (2014, p. 136) researched different variables, specifically for the implementation of an E-learning platform and with the aim of teaching SAP ERP (see Table 4.6).

Paa (2014, p. 127) surveyed with two groups for two years the implementation of an E-learning artefact for SAP ERP at the University of Innsbruck and collected the data through 570 questionnaires based on the variables on the Table 4.6. The duration of the course for the first group was 45 minutes, while the duration of the course for the second group took 90 minutes (split in two sessions of 45 minutes).

Table 4.5 Variables describing success of an information system (DeLone, & McLean, 2008, p. 238)

Dimension	Variables
System quality	Desirable characteristics of an information system: e.g., ease of use, system flexibility, system reliability, ease of learning, as well as system features of intuitiveness, sophistication, flexibility, and response times.
Information quality	Desirable characteristics of system outputs: e.g., relevance, understandability, accuracy, conciseness, completeness, understandability, currency, timeliness, and usability.
Service quality	Service quality refers to support that system users receive from the IS department and IT support personnel. For example: responsiveness, accuracy, reliability, technical competence, and empathy of the personnel staff.
Use	System use relates to the degree and way staff and customers utilize the capabilities of an information system. For example: amount of use, frequency of use, nature of use, appropriateness of use, extend of use, and purpose of use.
User satisfaction	This means the user level of satisfaction.
Net benefits	This approach the question to which extent Information Systems contributes to the success of individuals, groups, organisations, and industries. For example: improved productivity, improved decision-making, increased sales, cost reductions, etc.

Paa found (2014, p. 155) that the most important variables for a success implementation of an E-learning artefact in the first group were (as per order of importance): it runs without any application errors (system quality), it helps to prepare the work packages very well (net benefits), it is easy to use (system quality), it is quick in terms of answering the questions of the participants (service quality), the answers of the system administrator are understandable (service quality), the answers of the system administrator are helpful (service quality), it helps to prepare the tests in SAP very well (net benefits), the content is easy to understand (information quality), it is always available (system quality), and it provides a high content quality (information quality). Paa also found (2014, p. 156) for the second group with a course duration of 90 minutes similar results: it runs without any application errors (system quality), it is easy to use (system quality), it helps to prepare work packages very well (net benefits), the answers of the support are understandable (service quality), the answers of the support are helpful (service quality), it is always available (system quality), it helps users to learn whenever they want (net benefits), it helps to get started very efficient with the basics of SAP (net benefits), the information is free of contradictions (information quality), the answers of the system administrator are helpful (service quality).

Table 4.6 Variables for an e-learning artefact (Paa, 2014, p. 136)

Dimension	Variables
System quality	• The E-learning platform is always available • It's easy to use • It's runs without application errors
Information quality	• The E-learning tool provides a high content quality • Its content is easy to understand • Its content is provided through high quality media
Service quality	The service of system administrator: • allows an easy access to the E-learning platform • provides a professional support of the learning content • has the needed technical know-how to manage the E-learning • provides an enough degree of support online • replies to co-operative to improvement suggestions and proposals • replies in an understandable way • replies with answers that are helpful • replies in an amiable manner
Use	• The total time that users spent in the course • The times users accessed the E-learning platform • Users focus fully his attention to the E-learning platform • Users use the communications platforms of the E-learning tool regularly

(continued)

Table 4.6 (continued)

Dimension	Variables
User satisfaction	• The E-learning platform perception is very good • Also, for other type of learning is the E-learning platform as good complementary method • The personal learning success is very high • Users are very satisfied with the E-learning platform • Users like the E-learning experience • Users fully recommend the E-learning platform • Users believe others also may be very satisfied with the E-learning platform
Net benefits	• In general users see a big benefit with the E-learning tool • The E-learning platform provides a good basis to be able to work with SAP • It provides an overview in the SAP business processes • It provides an overview of the functions of an ERP system • It helps to get started very efficient with the basics of SAP • It allows users to learn whenever they want • It allows users to repeat the course as many times as they want • It allows to understand the principles of SAP in an efficient way • It motivates users to go through the content • It helps to prepare the work packages very well • It helps to prepare the tests in SAP very well • It provides a good understanding of business activities • It provides the needed skills to use SAP

Choi, Kim and Kim (2006, p. 223) developed a model to measure success of an ERP training with an e-learning platform. Starting from the flow theory in the context of e-learning domain the authors Choi, Kim and Kim (2006, p. 223) examine factors (see Table 4.7) that are relevant to positive learner's experience (*flow*) and preference (*attitude*).

Flow is defined as a holistic experience of total involvement towards individual engagement in the ERP training courses (Choi et al., 2007, p. 228). Attitude towards e-learning is determined by student's subjective probability of the consequence of a particular behaviour and influenced by student's evaluation towards the e-learning artefact (Choi et al., 2007, p. 228). They created a research model with defined factors as success measures for the flow and the attitude towards a e-learning artefact for ERP systems (Choi et al., 2007, p. 229). Choi et al. (2007, p. 235) model was conducted in 24 vocational schools with 960 distributed questionnaires, from which 236 were received and totally 223 were considered as valid. Choi et al. (2007, p. 238) found that:

Table 4.7   Factors and measures to measure success of an E-learning platform to teach ERP SAP (Choi et al, 2014, p. 240)

Relevant factors	Measures
Learner interface	• System is easy to use • System is user-friendly • System provides easy-to-understand content
Interaction	• Interaction with other students • Contact with instructor • Direct / timely feedback
Instructor's attitude towards students	• Friendliness towards students • Genuine interest in students • Advice and help
Instructor's technical competence	• Skills to explain how to use the related system • Care for students following up the ERP training • Skills to handle the related system
Content	• Content is up to date • Content fits learner's needs • Content is useful
Attitude towards e-learning	• Using the learning system is a wise idea • I like the idea of using the e-learning system • Using the e-learning system is pleasant
Flow experience	• Experience of flow • Frequency of flow • Intensity of flow
Technology self-efficacy in ERP system usage	• Mastered the use of ERP • Confidence to use ERP well • Ability to use ERP easily

- flow and attitude, as mediating variables, have a significant positive influence on learning outputs in terms of self-efficacy,
- students prefer an e-learning artefact for ERP if students are involved, concentrated, and showed intrinsic interests,
- the learner's interface of an e-learning artefact and content are important predictors for flow and attitude towards the e-learning artefact,
- and instructor technical competence has a significant influence on attitude towards learning but not on flow experience.

Even though the ISSM applied model and the flow model use different methods to identify the success factors for the effective learning of an ERP system, they have specific common requirements. There requirements refer to a well-structured and easy to understand content, the professional instructor technical competence and the possibility to interact with the teacher. On one hand, the flow mode of Choi et al. (2007, p. 239) was only conducted at high-schools and the response rate of a questionnaire was relatively low (23%). On the other hand, the ISSM model for the purpose of an e-learning artefact for ERP conducted by Paa shows a much wider analysis from different perspectives to help to identify the needed requirements for an e-learning artefact for ERP. Thus, the requirements found by Paa will be the base for the design of the e-learning artefact for SAP S/4 HANA.

4.5 Expositional Methods

Expositional methods refer to learning offerings, where the focus is the way the didactic materials in form of videos, text or audio are presented (Kerres, 2018, p. 330). The main objective of expositional methods is to guide the learner through learning materials based on a previously defined structure (Kerres, 2018, p. 330). According to Kerres (2018, p. 331) there are different methods in the literature upon which the learning digital structure can be built up. These are direct instruction, instructional events, cognitive master learning, adaptivity and learning analytics, and the 3–2–1 expositional model.

Rosenshine and Stevens (1986, p. 5) developed six fundamental functions for direct instruction. These six functions begin with a review of previous learned content, continues with a presentation of the new contents, guide the students practice, get feedback, and made corrections, allow independent student practice and make weekly as well as monthly reports. Similar to Rosenshine and Stevens, Gagné also (1985, p. 245) described the expositional structure of the learning through nine instructional events: demand attention, explain learning objectives, remember previous learning content, present learning material, provide learning help, apply the learning content, get feedback, test performance, and promote learning transfer to other areas. The cognitive master learning goes beyond having a structure, where the content of the materials is presented. It aims to bring learners into the working environment of a master, where they can experience first-hand how this master works, how does he plan his work, how does he evaluate his or her work (Kerres, 2018, p. 339). Lave and Wenger (1991, p. 52) apply this master experience to the cognitive level in terms of knowledge and competences. The idea behind is that the learners experience, through videos, the working environment of the master, and thus are able to make comments as well as discuss with other about the content of the videos (Kerres, 2018, p. 340). The adaptive and learning analytics focus on the idea that a learner should always be prepared for unexpected situations like disturbances or problems representing certain contents, which may vary the structure of the presentation (Kerres, 2018, p. 340). The teachers might e.g., realise during the learning that the presentation of the content is too difficult or too easy (Kerres, 2018, p. 340). In this kind of situations, the teachers could adapt the presentation of the content. In this sense the adaptation can be at a micro level (e.g., teaching or learning processes, or at a macro level (e.g., a complete adaptation of an entire course). Kerres (2018, p. 335) developed the instructional events model in to an expositional 3–2–1 model, and also grouped these instructional events into three element groups (see Figure 4.9). The first group represents the learning information, the learning material, and the learning exercises. While the learning information approaches the learning content, the learning materials structure should facilitate the learning of the processes (Kerres, 2018, p. 336).

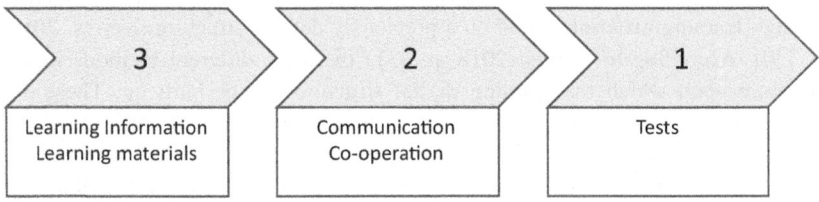

3	2	1
Learning Information Learning materials	Communication Co-operation	Tests

Figure 4.9 The 3–2–1 expositional model. Adapted from Kerres (2018, p. 336)

The third element refers to the test in the sense of checking and monitoring the learning results and learning success. The 3–2–1 expositional model approaches the idea that not all the elements for the representation of a structure is mandatory (Kerres, 2018, p. 335). For example, the group element Tests is less and less utilised due to more cognitive and constructionist applications of learning theories than behaviourist learning theories (Kerres, 2018, p. 335). In the cognitive and constructionist learning theories tests are less applied, while evaluation, benchmarking and certification are more significant.

Of all expositional models described in this chapter, the 3–2–1 expositional model will be part of the design cycle of the e-learning artefact. The reasons lay in the more comprehensive scope of the presentation model, as well as in the mandatory and facultative nature of the group elements. The next chapter shows the design cycle and operationalisation of DSR with the didactic methodical concept, the requirements of an e-learning artefact, and the 3–2–1 expositional model for teaching SAP S/4 HANA in the context of software services.

Design Cycle and Operationalisation of Design Science Research

5

The design of the e-learning artefact is based on an adapted and formalised operationalisation of DSR from Manson (2006, p. 163). According to Manson (2006, p. 163) this operationalisation counts with the steps awareness of the problem, suggestion, development, evaluation, and conclusion. Each of these steps has its own outputs. The outputs are defined through proposal, didactic concept, artefact, performance measures and results (see Figure 5.1).

5.1   Awareness of the Problem

The awareness of the problem results from an analysis in the practical world like in the context of an industry or a company, and also from the new technology developments, or from the reading in related disciplines (Manson, 2006, p. 162). The output of this awareness is a proposal. The awareness of this research project was given through the current issues in the area of knowledge transfer for SAP S/4 HANA. In the practical world it relates to the challenges that the company Leica Geosystems addresses in this area. Currently companies train their employees with Enterprise Resource Planning (ERP) systems in different ways, i.e., face-to-face training or using existing electronic learning (e-learning) platforms like simulations and gaming artefacts. E-learning can be defined as distance learning that creates and facilitates learning, anytime and anywhere, with the use of electronic media like internet, intranet, audio/video, CD-ROM as well as interactive TV (Choi et al. 2007, p. 224). These e-learning platforms focus on

Supplementary Information The online version contains supplementary material available at https://doi.org/10.1007/978-3-658-40731-5_5.

© The Author(s), under exclusive license to Springer Fachmedien Wiesbaden GmbH, part of Springer Nature 2023
F. Garayo Maiztegui, *Design and Evaluation of an E-Learning Artefact for the Implementation of SAP S/4 Hana®*, Gabler Theses, https://doi.org/10.1007/978-3-658-40731-5_5

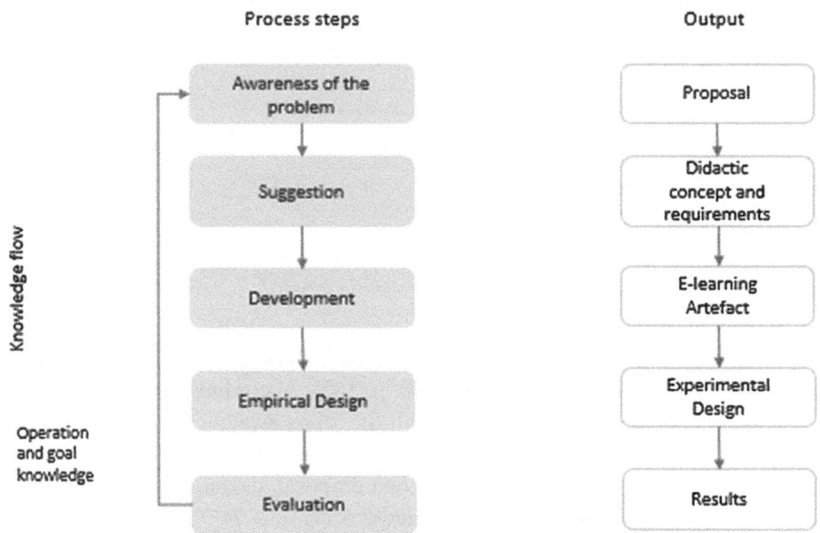

Figure 5.1 Outputs of design science research (adapted from Manson, 2006, p. 163)

standard business processes represented in the ERP systems. However, these ERP e-learning platforms do not satisfy many companies' needs. The reason for this is the continuous customization and improvement of business processes, as well as the consequent and permanent training of employees. If no proper training of employees does happen, there will be also errors in the order entry with subsequent customers complains. There has been some research in terms of e-learning simulation trainings for standard ERP business processes, however little research has happened in the design and implementation of e-learning artefacts for ERP systems considering customised business processes in the context of companies' environments. This specially plays an important role when Fiori Apps for SAP S/4 HANA are developed for customised business processes.

Additionally, a literature review was carried out to identify if already e-learning solutions in the area of SAP ERP exist. The review question of the literature review needs to be described very precisely because this will underpin the decision of including or excluding scientific papers (Jesson, Matheson, & Macey, 2011, p. 110). The review question is formulated as follows: what empirical evidence is available in the implementation of e-learning artefacts for the knowledge transfer of business processes for Enterprise Resource Planning

(ERP) systems in adult education? The additional following sub questions will be addressed: is the nature of the evidence qualitative or quantitative, in which countries are the studies set in and which research designs are used to generate evidence (Jesson, Matheson, & Macey, 2011, p. 111). The literature review contains a methodology, a search strategy, the results of the search strategy, a quality assessment and synthesis following a concept centric approach.

According to Rowe (2014, p. 243) there are different types or reviews: descriptive reviews, reviews for understanding and explaining, and reviews that focus on testing. Descriptive reviews normally adopt a systematic methodological approach, summarise research using categories, look for areas of research that might be beneficial, are normally atheoretical (Schryen, Benlian, Rowe, Gregor, Larsen, Petter, Paré, Wagner, Haag, and Yasasin, 2017, p. 761), and focus on empirical literature that has been produced on the research topic (Rowe, 2014, p. 244). Reviews focused on understanding and explaining are normally narrative, aim to understand a new phenomenon or problem, and contribute to theory building (Rowe, 2014, p. 244). Reviews that focus on testing are mostly meta-analysis that gather empirical evidence, as well as aggregate effect sizes (Schryen et al. 2017, p. 761). While meta-analysis reviews comprehend a set of statistical methods for combining quantitative results, systematic methodological reviews acknowledge critically former research data using organised, transparent, and replicable search procedures (Littell, Corcoran, & Pillai, 2008, p. 1). The following literature review is descriptive, because it aims to summarise research knowledge based on a concept centric approach, as well as identify areas of research that can benefit for further research. The methodology of this literature review is also systematic. Systematic means that the review is taken according to a method or a fixed plan (Gough, Oliver, & Thomas, 2012, p. 5), and it requires a rigorous and exhaustive search for gathering and synthetising the findings of studies about a particular topic or question (Jesson et al., 2011, p. 104). A systematic literature review comprehends used secondary studies to find, evaluate, and aggregate relevant primary studies on a specific research topic (Dresch, Pacheco-Lacerda, & Valle-Antunes, 2015, p. 129). Dresch et al. (2015, p. 132) as well as Gough et al. (2012, p. 8) propose different steps when carrying out a systematic literature review. These steps are search strategy and the quality assessment. The search strategy tries to answer questions like what to search for, where to search, how to minimise bias, what studies should be considered and the extend of the search (Dresch et al.,2015, p. 136). Additionally, the search strategy also includes the way the documentation gets included to make the decision transparent in the sense that other researchers can replicate it (Jesson et al., 2011, p. 111).

For this, following information will be needed: the title of the database or scientific journal, data searches conducted, years covered, search terms, language restrictions and number of hits (Jesson et al., 2011, p. 111). The search strategy aims to find the largest possible number of relevant studies (Jesson et al., 2011, p. 137). The search strategy considers also different dimensions (see Figure 5.2): the search terms, the search extent, and the resources (Jesson et al., 2011, p. 137). The search terms are the first approach to be addressed in terms of what to search for, while the search of resources refer to where to search (Jesson et al., 2011, p. 136). Further, and based on the review question, inclusion, and exclusion criteria was considered (Jesson et al., 2011, p. 137). The search terms (see Table 5.1) for this literature review will consider the English language. The reason for this, is that e-learning as well as ERP (Enterprise Resource Planning) are standard foreign words used in several languages (e.g., in German or Spanish). Further, most of the scientific papers in foreign languages commonly contain an abstract in English, that contains the topic, the thesis, and the frame results. The objective of the selection of the search terms is to accomplish in a most comprehensive manner a way to find the largest possible number of relevant studies (Dresch et al., 2015, p. 137). The operators used will be AND, OR, "", and— (minus). The reason to use a minus is to exclude the medical studies from two scientists with the surname "van ERP". This is especially crucial in the scholar google search. The search sources for the research are related to three types: the Karlsruhe Virtual Katalog (KVK) library catalogue, the databases Web of Science, ScienceDirect, Research Gate, ProQuest and ABI/Inform (ProQuest), as well as several relevant journals and conferences in the area of education, computer science, e-learning, and information systems. The library catalogue relates KVK has access to 80 bibliographical databases, with 1.854.962 printed books, 183.602 e-books, 1089 journals, 166.433 electronic journals and 175.132 university journals. While EBSCOhost includes 375 full and secondary databases with more than 420000 e-books and 355000 electronic journals as well as e-journals packages, Web of Science is a multi-disciplinary database with more 12000 of the highest impact journals worldwide, as well as 150000 conference proceedings (Dresch et al., 2015, p. 141). Science Direct contains more than 21000 titles and 50 million records (Dresch et al., 2015, p. 141). ABI/Inform (ProQuest) contains more than 450000 working papers, 10000 business cases and over 22000 PhD Dissertations selected from its Dissertations and Theses database (Dresch et al., 2015, p. 141).

Additionally, the search also includes a manual approach in different relevant journals and conference papers in the area of education, computer science, e-learning, and information systems. The inclusion criteria for the selection of the

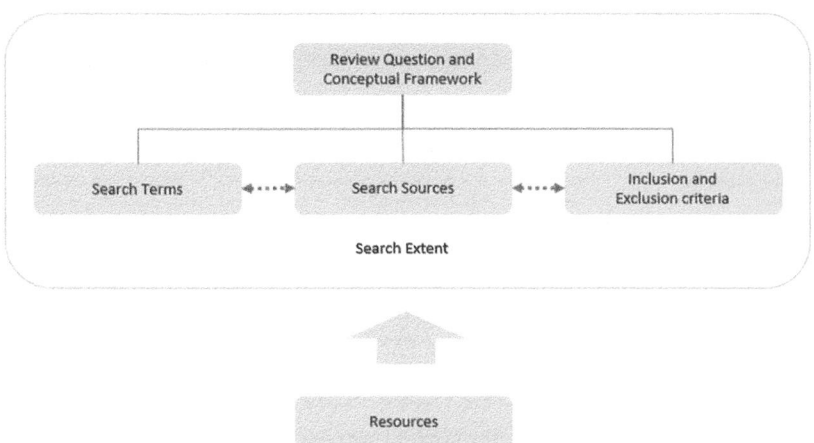

Figure 5.2 Search strategy (Dresch et al., 2015, p. 137)

Table 5.1 Search terms. (Own creation)

Search term Aspect 1	Search terms Aspect 2	Search terms and Operators (AND, OR, "", -)	Location	Year
ERP	e-learning, elearning	ERP AND "e-learning" AND (teaching OR learning OR training) AND (online OR internet OR web) - "van ERP"	In title and in text	2000-2022 (Year to Date)
	teaching			
	learning	ERP AND "elearning" AND (teaching OR learning OR training) AND (online OR internet OR web) - "van ERP"	In title and in text	2000-2022 (Year to Date)
	training			
	online			
	internet	ERP AND (learning OR teaching OR training) AND (online OR internet OR web) - "van ERP"		
	web			

articles and decision-making process consider four steps (see Table 5.2). The first step is based on the findings of the search terms in the title and on the text of the journal article or dissertation. In the case of Google Scholar, the title will be checked first. The reason for this, is to reduce the number of thousands of articles that are not related to the topic (e.g., event related potential). For the first step the inclusion criteria include studies that are related to the ERP training, ERP

Table 5.2 Inclusion criteria. (Own creation)

Inclusion scope	Relevance
Countries	All countries.
Publication type	Peer-reviewed papers in academic Journals, dissertations, and publications in academic conferences.
Quality of research	Quantitative and qualitative assessment according to McNeill and Chapman (2005).
Analysis	*First step*: search terms in the title and in the text. *Second step*: backward and forward search approach for articles found in step 1. *Third step:* implementation of e-learning artefacts as well as empirical qualitative, quantitative research papers. *Fourth step:* analyse if peer reviewed and relevant journals, as well as dissertations.
Research designs	Quantitative longitudinal and cross-sectional research, Qualitative, Mix-methods.
Time horizon	From year January 2000 till January 2022
Search fields	Business, management, computer science, information systems, education, e-learning.
Language	English.

learning or ERP teaching in the title or in the text. The second step relates to a backward and forward approach (Webster, J., Watson, R.T., 2002, p. XV) of articles found in step 1. A backward approach aims to review articles identified in the reference list in step 1 to find out prior articles, while a forward approach tries to identify articles using the citing electronic version of the citation index in the Web of Science database (Webster, J., Watson, R.T., 2002, p. XV). The third step comprehends the reading of the articles and dissertations to identify studies related to the implementation of e-learning systems, as well as empirical qualitative, quantitative and interventionist research in the context of ERP systems. The fourth step checks if the articles are peer-reviewed and published in relevant journals (see below rating). Although the experiment of the research project is quantitative, also qualitative papers were taken into consideration. There could exist e.g., interviews with experts, whose results hint the benefits of an e-learning artefact for ERP teaching and learning purposes. The journals where the articles are published, will be analysed in terms of peer-review and the journal rating according to the "Verband der Hochschullehrer für Betriebswirtschaft e.V.", the Journal Impact Indicator (JIF) from the Web of Science, the Scimago Journal and Country Rank (SJR) as well as the Source Normalized Impact per Paper (SNIP) from Scopus. Please see below the meaning of the rating A +, A, B and C

(Verband der Hochschullehrer für Betriebswirtschaft e.V., 2021, „VHB-Jourqual, VHB-Jourqual 3", para. 3):

- A + = outstanding and worldwide leading scientific journals in the field of business administration,
- A = leading academic business journals,
- B = important and respected scientific business studies journals
- C = recognized academic business journals

The Source Normalised Impact per paper (SNIP) measures citations weighted by the subject field, where a SNIP higher than 1 means more citation than average, and a SNIP of 1.5 generally indicates a well-cited journal (Massey University, 2021, Using Scopus and SJR to find a Journal's Impact and Rank, para. 1). The Scimago Journal and Country Rank (SJR) measures weighted citations received in specific year to documents published in the journal in previous three years (Scimago Journal and Country Rank, 2021, Help, Understanding indicators, tables and charts, para. 1) A SJR higher than 1 means above average citation potential (Nottingam Trent university, 2021. p. 6). The Journal Impact Factor (JIF) is the *"number of citations received in a given year, to a journal's previous two years of publications (linked to the journal, but not necessarily to specific publications) divided by the sum of citable publications in the previous two years"* (Elsevier, 2021, Measuring a journals impact, para. 10); e.g., an article in a specific year with a Journal Impact Factor of 2, on average, it has been cited two times.

The assessment of the quality research is the next important appraisal of the material selected (Jasson et al., 2011, p. 116). According to Petticrew and Roberts (2006, p. 127) quality research refers to the reliability, internal validity (a study is free of methodological bias) and external validity (the possibility to generalise the results of the trial to other settings). According to McNeill and Chapman (2005, p. 9), a questionnaire as a method is reliable, if anyone using that method, or the same person using it at another time, would come up with the same results. Validity refers to the question if the data collected is a true picture of what is being study (McNeill, & Chapman, 2005, p. 9). The quality assessment can be based on qualitative and quantitative studies. For the qualitative studies the survey method in terms of a pragmatic approach is considered, and it is based on the details given by the authors of each article about the survey design (McNeill, & Chapman, 2005, p. 117). On the other side the assessment of quantitative studies is based on following criteria: the objective of the study, the theories the research is linked to, the research design, the research method and sample size, the reliability as well as internal validity.

This literature review considers the articles found according to the methodology in terms of search strategy and quality assessment. It also includes a concept centric synthesis. The search strategy begins with the search of articles with the help of the search terms and the search sources. Different source databases contain different subjects of scientific research. The selection of appropriate subjects that are related with the research topic could make the searching more focused towards ERP in terms of Enterprise Resource Planning.

Also, a manual search was done in selected journals in the area of Information Systems, Computer Science, Education and E-Learning. After searching for papers based on the search terms through the search sources, 33891 hits were found (see Figure 5.3). Of all these hits, in the first step 102 articles related to any online, web, internet, e-learning as well as of ERP training, ERP learning, or ERP teaching could be found. As a second step, a backward and forward approach was taken based on these 102 articles. For the backward approach 12 new articles in the context online, web, internet, e-learning as well as of ERP training, ERP learning or ERP teaching of could be found. For the forward approach only existing ones from previous searches could be found. This makes a total of 114 articles. As a third step the abstract and the text was read in terms of empirical studies referred to an implementation of e-learning tools as well as empirical qualitative, quantitative and interventionist research, where 53 research papers could be identified. For the fourth step, after checking if they were peer reviewed as well as considering the relevance of the journals, 28 research papers could be selected. For the quality assessment (see Appendix A in additional electronic material) the following criteria was considered: the objective of the study, the theories the research is linked to, the research design and the research method, the sample size, the reliability as well as internal validity. The studies S2, S6, S12, S14, S20, S22 and S26 do not link their research to any research theory or model. This means that their results might have a relevance for the practical world, however, it does not provide a knowledge contribution for the scientific world. Also, the studies S1, S2, S6, S7, S8, S9, S10, S11, S12, S14, S15, S17, S19, S20, S21, S22 and S26 do not contain any information related to the internal validity in their studies. This is relevant to see if the studies are free of methodological bias. According to Weber and Watson (2002, p. XVI), a literature review should be concept centric. The authors also recommend compiling a concept matrix (see Appendix B in additional electronic material) as well as synthesise the literature by discussing each theme (Weber, & Watson, 2002, p. XVII). For the synthesis of the literature review following themes could be defined: knowledge, learning, skills, performance, and technology (see Table 5.3). Table 5.3 shows that

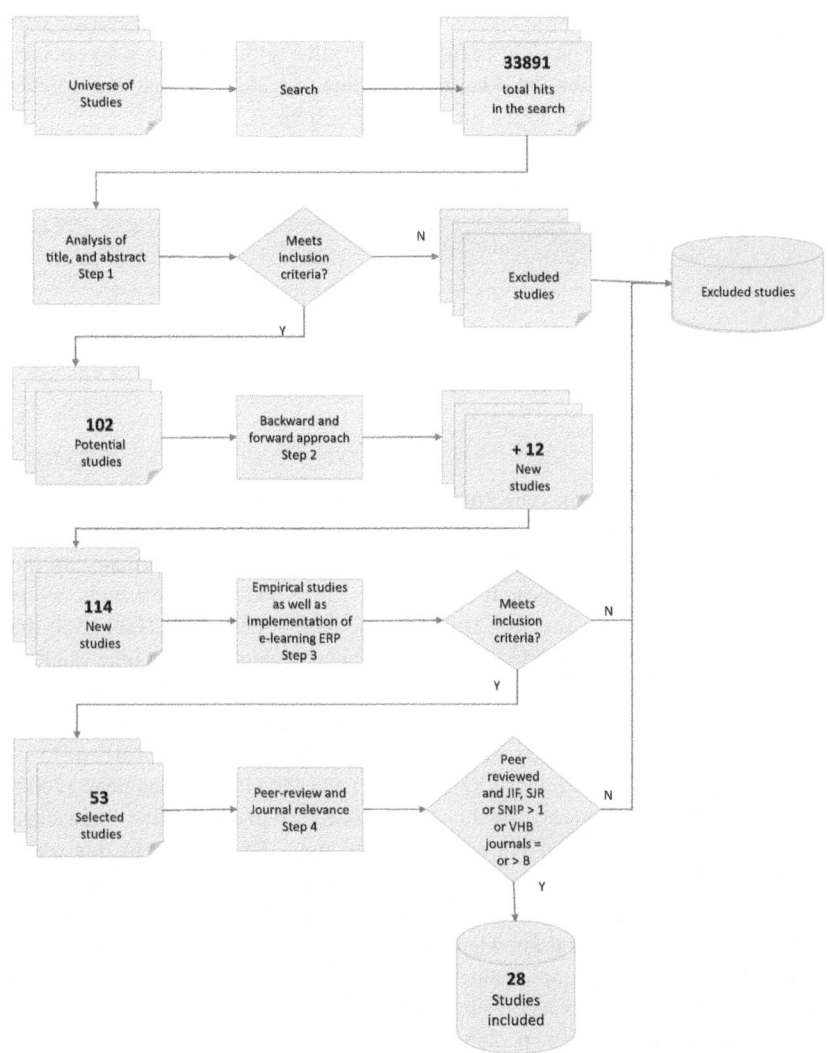

Figure 5.3 Search process and eligibility numbers (Dresch et al., 2015, p. 143)

researchers have utilised several research design and methods in measuring outcomes like knowledge, learning, performance, skills, and technology acceptance at the time of implementing an e-learning artefact or using an e-learning artefact for other research purposes in the area of ERP systems.

Table 5.3 Themes, research designs and methods. (Own creation)

| | | Quantitative Research Designs | | | | | | Qualitative Research Designs | Mix-Methods Research Designs | |
| | | Quantitative Longitudinal research | | Quantitative Cross-sectional research | | | | Qualitative research | Quantitative Cross-sectional research | Qualitative research |
Themes	Research methods	Experiment	Survey	Case study	Experiment	Quasi-Experiment	Survey	Survey	Experiment	Survey
Knowledge		1	2	1	1	1	2	1	2	1
Learning			3	2	2		4		1	1
Performance					2	1				
Skills		1		1	1		7	1	1	
Technology acceptance			3		1	1	7			

A common approach in nine studies is the measurement of knowledge. The choice of the research methods includes for the longitudinal and cross-sectional research designs experiment, quasi-experiment, survey, and case studies. In the case of mix-methods, the research adopts experiment and survey methods, while in the case of qualitative research it includes a survey method. In Table 5.3, the numbers related to the knowledge, for example, are higher than nine studies. This is because in one of these studies more than one research method applies. Knowledge is a broad theme and can be understood from different perspectives, as well as be classified in different types (Alavi, & Leidner, 2001, p. 113). According to Alavi and Leidner (2001, p. 113) and Gravill and Compeau (2008, p. 289) the theme knowledge in this synthesis refers to declarative knowledge (know-what) and procedural knowledge (know-how). Knowledge in this sense is understood from several perspectives: e.g., enterprise management knowledge (declarative knowledge), business process knowledge (procedural knowledge) and ERP knowledge (procedural knowledge). Of nine articles, six of them have as an independent variable the implementation of an e-learning artefact, while three of them have other type of independent variables like ERP self-regulated strategies (Gravill, & Compeau, 2015, p. 288), team collaboration effectiveness (Darban, Kwak, Deng, Srite, &, Lee, 2016, p. 98), and mental rehearsal (Shah, 2015, p. 74). The dominant idea in the six studies, is that the implementation of an e-learning artefact positively influences the acquisition of business process knowledge. Cronan and Douglas (2012, p. 3) find out through a longitudinal pre-post-test study also, that an ERP simulation program significant increase students' knowledge about business processes. In a e-learning simulation game, operational business processes are put into practice through operational tasks that are shared between team players (Cronan, Léger, Robert, Babin, & Charland, 2014, p. 464). Upon these operational tasks, decisions are made. This enhances the business process knowledge of participants through continuous interaction among the team players considerably. However, also business process knowledge can be acquired at an individual level. For example, Körsgen (2000, p. 252) created a Computer Based Training program for ERP systems and identified that the individual training of different ERP modules allows students to understand and acquire business process knowledge. Contrary to the positive mainstream of e-learning artefact and ERP business process knowledge, the studies of Monk, Guidry, Pusecker, & Ilvento (2018, p. 98), and Pakinee and Purinat (2021, p. 4063) show other results. In these studies, the implementation of an e-learning artefact does not have a positive influence in the acquisition of knowledge. The reasons behind are the positive as well as negative influence of learning approaches depending on different learning types. For example, on one side Monk et al. (2018, p. 98) affirm that e-learning training was better perceived by some learners due to the availability of the material online, as well as learning through quizzes. However, on the other side students were exposed to more distractions at home. Both studies approach the research with an experiment as research method, where a gamified and non-gamified e-learning artefact design, as well as a face-to-face training and a computer assisted pedagogy are compared. In another study Gravill and Compeau (2008, p. 289) find that self-regulated strategies positively influence declarative knowledge (know-what) and procedural knowledge (know-how). Self-regulated strategies are influenced by the subjective opinions of learners about their own learning needs. Depending on how learners define their own learning strategies, the result of the knowledge as an outcome will be different. In one further study Darban et al. (2016, p. 98) confirm team collaboration determines individual effort, as well as perceived knowledge update. This is crucial for game designers to identify which team collaborations features should be considered at the time of developing e-learning artefacts. The e-learning artefact in most of these studies (7 of them) are based on an ERP simulation game called ERPSim from the ERP provider SAP. This is a simulation program that allows the learners to focus on analytics and decision-making by incentivising them in order to implement strategies using the ERP system (ERPSim Lab, the Business Simulation for SAP S/4HANA, para. 2). However, participants do not learn how to perform a specific business process (e.g., Subscription software sales process) of a company, and thus the ERPSim does not cover many times the needs to perform companies' standard business process steps in an ERP system. Further research is needed to understand why companies opt not to participate in the e-learning simulation game (Deranek, McLeod, & Schmidt, 2019, p. 380). Also, future research mentioned in these studies is the need to define objective measures in order to assess acquired knowledge.

Another important theme as a result of the categorisation is the theme learning as the process of acquisition of knowledge. Common to nine studies is the measurement of the variable ERP learning at the time of implementing an electronic ERP simulation game. The variable learning is approached from different perspectives. For example, studies from Cronan and Douglas (2011, p. 231), Seethamraju (2011, p. 19) and Heričko et al. (2017, p. 139) try to measure the learning effectiveness. Cronan and Douglas (2011, p. 231) summarise learning effectiveness into simulation experiences, ERP learning, an attitude. Pakinee and Purinat (2021, p. 4063), Alcivar and Abad (2015, p. 117) look to the theme learning in the sense of ERP learning itself. Also, Heričko et al. (2017, p. 139) consider that the effectiveness of the learning process consists of three components: the ERP application, the business context of the ERP application and the collaborative tasks enabled by the ERP application. Additionally, Heričko et al. (2017, p. 130) postulates the experimental learning cycle as a three-step approach in terms of understand (how to play the game), act (take decisions) and reflect (reflect on feedback). The dominant idea in the findings of these articles is that the implementation of an electronic ERP simulation game has a positive influence in the learning effectiveness as well as the ERP learning itself. Also, pre-, and post-simulation results show that students had a positive learning experience with the ERP simulation game (Cronan, & Douglas, 2011, p. 231), while the ERP simulation game also enhances the process understanding of the learners and thus reinforce its pedagogical effectiveness (Seethamraju, 2011, p. 26). Learning with an ERP simulation game happens in teams that need to collaborate in order to fulfil the tasks assigned by the system. However, for the enhancement a lecturer or trainer is needed to explain the guidelines of the simulation game (Pakinee, & Purinat, 2021, p. 4063). One study looks at the learning from the perspective of learning value in terms of attitude, perception on learning, and pedagogical preference (Angolia, & Pagliari, 2018, p. 119). Angolia and Pagliari (2018, p. 119) find that students participating in an extended ERP simulation game have more perceptual learning value, even though they had to invest more time in preparation and also had more time pressure. Another study (Li, & Tsai, 2020, p. 4) find that mastery goal orientation has a positive influence on learning outcomes in the sense of learning satisfaction and learning motivation. This positive influence is given even though motivations behind these goal orientations are sometimes different (Li, & Tsai, 2020, p. 15). According to the goal orientation theory, a mastery goal orientation will show adaptive pattern responses, when students confront with new challenges (Li, & Tsai, 2020, p. 6). These adaptive patterns cause, through self-confidence, a positive influence with higher levels of learning satisfaction and motivation (Li, & Tsai, 2020, p. 6). In this area of learning further research is needed in terms that these simulations would need to include customising business processes of existing companies with real examples (Alcivar, & Abad, 2015, p. 118). Also, there is a need to create a new e-learning prototype and adjust game elements, as well as make them more suitable for learner personality traits (Pakinee, & Purinat, 2021, p. 4065).

Further, a comparison of traditional methods of ERP classrooms with e-learning artefacts in form of an experiment would be meaningful and important for the research (Cronan, & Douglas, 2012, p. 10). Future research could also examine the learning effectiveness of training on employees in a corporation (Shah, 2015, p. 112). Three studies measure performance when implementing an electronic ERP simulation game. Common to all studies is that the measurement of performance is based on objective measures like tests and exercises. While Parush, Hamm, & Shtub (2002, p. 323) measures performance with the variables profit, due date performance (sales orders in the system supplied on time) and run duration (elapsed time needed to run each simulation), Alcivar and Abad (2015, p. 112) measures performance in terms of the results achieved in the tests and exercises completed during the training. Conroy (2012, p. 3) also measures performance in terms of exams. According to Parush et al. (2002, p. 319), as well as Alcivar and Abad (2015, p. 109), the implementation of an electronic ERP simulation game has a positive influence in the performance during the learning process. In the study of Parush et al. (2002, p. 328) performance was measured using an experiment with a control group and a treatment group. While the treatment group could review the learning history in the e-learning artefact, the control group could not. The results show a better performance of the treatment group for the profit and due date performance. In the study of Alcivar and Abad (2015, p. 112) the authors confirm that this positive influence on performance is given because students perceive the ERP simulation game as enjoyable. Contrary to these results Conroy (2012, p. 65) finds in his experiment with a treatment group and a control group that there is not a performance asset when using an ERP simulation game in comparison to not be using it for ERP training purposes. Unfortunately, the sample size of this dissertation due to the complex nature of the research is with 16 employees too low. That is the reason why the author recommends a higher sample size. In order to measure performance further research is needed to utilise customised processes from companies that can enable a more realistic contextualisation of the content of ERP trainings (Alcivar, & Abad, 2015,

p. 118). Parush et al. (2002, p. 331) emphasised the need of further research in the process of learning and utilising the e-learning artefact in terms of visualising the learning history, as well as recording the history automatically or manually. Both aspects might have an implication in the learning process and thus in the performance.

Another theme relevant for the synthesis of the literature review are the skills. For this theme nine studies could be consolidated. Three studies try to measure skills as a dependant variable with the implementation of an e-learning arte-fact as an independent variable. Skills is understood as the needed ERP technical skills gained through the participation on an electronic ERP simulation game. The common idea in these studies is that the implementation of an electronic ERP simulation game contributes beneficially to the enhancement of learners' ERP skills. For example, the pre-post-test study of Cronan and Douglas (2012, p. 9) identify that the ERP simulation game positively influences ERP skills, where students gain the needed ERP competencies to perform tasks. Also, the survey of Seethamraju (2011, p. 19) brings up that an ERP simulation game significantly improve their process orientation and integrative skills. Further, the survey of Heričko et al. (2017, p. 130) come up with similar results saying that students perceive to have sufficient ERP skills to perform the needed ERP tasks. In additional studies, skills are defined as self-regulatory skills in the sense of thinking skills and process orientation (Monk et al. 2018, p. 97). While ERP skills allow learners to have the needed competences to perform the tasks, thinking skills enable them with the ability to take decisions. The experiment and semi-structured interviews carried out in the study of Monk et al. (2018, p. 97) confirm that the enhancement of self-regulatory skills of learners through an e-learning artefact are evident. There are also other studies (four studies) that try to see the influence of an e-learning artefact in the self-efficacy as a dependent variable. Self-efficacy is understood in these studies as the subjective ability in terms of skills to use a computer system (Scott and Walczak, 2009, p. 221). However, the independent variables of these studies are conditioned either by factors of the e-learning success model (Choi et al. 2006, p. 223), by self-regulated learning strategies (Gravill, & Compeau, 2007, p. 289), by ERP knowledge mastery (Mullins, & Cronan, 2021, p. 1), or by computer self-efficacy antecedents (Scott and Walczak, 2009, p. 221). These research projects utilise the e-learning artefact for the empirical validation of self-efficacy. Thus, for example, Gravill and Compeau (2007, p. 293) find that there is a strong positive relationship between effective use of online learning strategies and self-efficacy. This is the case because individuals learn on their own in different ways in terms of applying learning strategies to fulfil their own needs (Gravill, & Compeau, 2007, p. 293). While Mullins and

Cronan (2021, p. 8) state that ERP Knowledge positively influences self-efficacy, Scott and Walczak (2009, p. 228) find that organisational support and engagement are statistically significant computer self-efficacy determinants. Choi et al. (2006, p. 223) also find that flow experience has a direct impact on self-efficacy that it is needed to perform tasks using ERP systems. These results in the studies of self-efficacy show, how different factors and antecedents have a positive influence in the acquisition of self-efficacy competences. On other adjacent study to the theme skills, research focuses on the prediction of ERP Knowledge towards ERP competency (Charland, Cronan, Léger, & Robert, 2015, p. 31). This ERP competency refers as well to the needed ERP skills to perform tasks by the learners. The authors make a difference between basic and complex existing knowledge of an ERP system and find out that basic knowledge is important to develop ERP competency. However, complex knowledge is not a predictor of gained competences, when it comes to solve problems using an ERP system (Charland et al., 2015, p. 31). Monk et al. (2018, p. 97) state the challenge of generalising the applicability of learning effectiveness to the graduate skills, especially when it comes to apply them in a company workplace environment. Heričko et al. (2017, p. 131) affirm that further scientific gaps still exist in the area of how to increase retention of less understood SAP technical transactions, as well as about advantages of using electronic ERP simulation games with regard to other skills like teamwork. According to Mullins and Cronan (2021, p. 10) also future research of interest could be to investigate across geographies and cultures, and in different types of organizations implementing ERP systems.

For the technology acceptance theme several determinants according to the Technology acceptance Model (TAM) of Davis, Bagozzi, & Warshaw (1989) are mentioned. A common approach in four studies is the measurement of the variable attitude. These studies understand attitude in terms of a psychological commitment of users towards ERP systems (Deranek et al. 2019, p. 374). For example, Cronan et al. (2011, p. 231) find that attitudes before and after the electronic simulation game are very high and that the variable ease of use increases significantly after the electronic simulation. Also, Deranek at al. (2019, p. 373) as well as Hwang and Cruthirds (2017, p. 60) state positive influence of the attitude of learners towards ERP with the implementation of an electronic simulation game. Mullings and Cronan (2021, p. 1) state on the other way round that an ERP knowledge is an important antecedent of attitude. The result of this study shows an interdependency between ERP knowledge and technology acceptance. Additionally, three articles study the influence of the implementation of an electronic ERP simulation game towards perceived usefulness, and perceived ease of use (also determinants of the TAM model). For example, Chauhan et al., find that certain determinants like performance expectancy, and effort expectancy positively impact students' behavioural intention to use. The survey of Deranek et al. (2019, p. 298) also affirms that the results show that the use of media content (like videos and audio tutorials) could make the learning experience more useful and acceptable to the user. The studies of Schotz and Kapeso (2014, p. 287) and Babaian, Xu, & Lucas (2017, p. 200) state that an e-learning artefact positively influence the perceived usefulness and perceived ease of use. One study (Conroy, 2012, p. 65) looked to the variable acceptance from the perspective of affective psychological state. He finds that the application of an ERP simulation game does not have an impact on the affective outcomes of the training (Conroy, 2012, p. 3). In this study unfortunately the sample size of the experiment was with 16 participants due to the complexity nature of the study too low. Another study from Mullins and Cronan (2021, p. 1) affirms that knowledge is an important antecedent of perceived ease of use and perceived usefulness. In this sense it can be identified that there is an interdependency between knowledge and technology acceptance. Factors might determine the business process knowledge or ERP knowledge of the user, but also on the other way round ERP knowledge affects positively the technology acceptance of the user. Also, Karaali, Gumussoy, & Calisir (2011, p. 343) analyse factors affecting the usage of a web-based mode and find that perceived usefulness, attitude toward use, and social influence were found to be significant predictors of the behavioural intention to use web-based learning system for an ERP system (Karaali, Gumussoy, Calisir, 2011, p. 350). In terms of research Deranek et al. (2019, p. 299) state that further investigation is needed to explore aspects like content quality (e.g., the length and quality of the videos). Future research is also needed when it comes to find objective measures in order to measure knowledge enhancement with the implementation of an electronic ERP simulation game.

As a summary it can be said that a total 28 relevant studies could be found. Out of these 28 studies, 18 defined the implementation of the e-learning artefact as an independent variable (15 studies were based on an electronic ERP simulation game and three cases on an own developed e-learning artefact). The other remaining 10 studies defined other independent variables (i.e., computer self-efficacy determinants, self-regulated strategies or factors affecting the decision of using a web-based model) and utilised an e-learning artefact for this purpose. All articles could, however, be consolidated in the following themes: knowledge, learning, performance, skills, and technology acceptance. Table 5.4 provides an overview about quality assessment, the concept centric themes, and also the major findings, limitations, gaps, and the contribution of the research project to address these gaps. From 28 studies, 9 studies do not link their research with a theory

Table 5.4 Overview of learnings. (Own creation)

Quality Assessment	• Studies S2, S6, S12, S14, S20, S22 and S26 do not link their research to any research theory or model. This means that their results might have a relevance for the practical world, however, it does not provide a knowledge contribution for the scientific world. • Also, the studies S1, S2, S6, S7, S8, S9, S10, S11, S12, S14, S15, S17, S19, S20, S21, S22 and S26 do not contain any information related to the internal validity in their studies.
Themes	• Knowledge, Learning, Performance, Skills and Technology acceptance
Major findings	• The majority of the studies show that the application of an e-learning artefact has a positive and beneficial influence in the knowledge of business processes and ERP knowledge • Studies show that the implementation of an e-learning artefact had a positive influence in the ERP learning • Further the performance was also positively influenced by the implementation of an e-learning artefact • There was also a common and positive influence in the variable of ERP skills of all studies at the time implementation of an e-learning artefact • Technology acceptance in terms of attitude was in all studies positively influenced by implementing an e-learning artefact
Limitations	• Most of the studies were done at universities with students. The only two studies that were done at companies have a small sample size (52 and 16 employees). Such studies carried out in a high school or in companies may vary the outcomes and learning self-efficacy considerably due to different learning environments as well as the age of the individuals (Chu, 2010, p. 255) • The majority of the studies are based on perceptual data as well as self-reported measures, and not in objective grades and test performance • Information overload of the ERP simulation systems • Research is made in only in one country (examples were studies only made in Japan, or in India or in Japan) • Using the simulation with experienced employees, familiar with information systems, may cause that the constructed of the technology acceptance need to be redefined.
Gaps	• Lack of transferring business process knowledge with ERP simulation games when it comes to learn about customised business processes at companies. • There is little research effort for developing a theoretical model and verifying it with an empirical study, especially within the context of an ERP e-learning system • Future research should examine the effectiveness of training on employees in a corporation. Also, future research should evaluate whether the results obtained cover firms in other countries. • More strategic sampling and data gathering methods with a more diversified sampling is needed • Understand why companies opt not to participate in an ERP simulation system • How to increase retention of less understood SAP technical transactions
Contribution of the research project to address the gaps	• The content of the e-learning represents real business process case of a company representing the sales of a software subscription. • The study develops a theoretical method with a prototype and aims to validate it win an experiment as empirical study • The sample size should be big enough to be representative • It provides a contribution to the literature as well as to the scientific world

or model. Also, in 17 studies there is no information related to the validity in their studies. This is relevant to see if the studies are free of methodological bias. The studies were synthetised using a concept centric approach with categories, where also the types of variables and gaps in the literature were identified. Major findings of the studies show that the application of an e-learning artefact has a positive and beneficial influence in the knowledge of business processes and ERP knowledge. In these studies, the research methods of experiment, survey and case study were applied. For the learning theme there are common findings in all the studies. In this sense it can be said that in all these studies the implementation of an e-learning artefact had a positive influence in the ERP learning. Additionally, the learning effectiveness, the experience of learning and the learning value was also beneficial. Further, the performance was also positively influenced by the implementation of an e-learning artefact. The performance was objectively measured with tests and exercises. Only one study (Conroy, 2012) did not find any difference, however he recommended a study with a bigger sample size. There was also a common and positive influence in the variable of ERP skills at the time of implementing an e-learning artefact. Finally, it can be said, that the variable technology acceptance in terms of attitude was in all studies positively influenced. Common limitations were also found in some studies. For example, most of the studies were done at universities with students. Studies carried out in a high school or in companies may vary the outcomes and learning self-efficacy considerably due to different learning environments as well as the age of the individuals (Chu, 2010, p. 255). Unfortunately, the only two studies that were validated at companies had a quite small sample (16 as well as 52 adults). Also, these studies were carried out in one single country, which limits the experiment into one culture. The dominant idea is also that the surveys mostly are based on the collection of data through questionnaires, which makes the studies to be based on subjective perceptual data, and less on objective measures like tests and grades. Also, gaps could be identified in the literature. Especially remarkable was that the studies did not consider customised business processes of companies, and that the studies were mainly not carried out in companies. This could be of interest because the work environments as well as the average age of the participants may provide a different outcome. This makes the generalisation of research projects for adults in enterprises environments difficult (Chua, 2010, p. 255). In this sense the effectiveness of an e-learning artefact in the context of a company is equally missing. Further, there is very little research available in terms of developing a new model, as well as validating an e-learning artefact in the context of a company. In the studies it could also be identified that more data gathering methods, as well as a bigger sample is needed. Additionally, further scientific gaps in the area of how to increase retention of less understood SAP technical transactions should be approached. The planned research addresses these gaps in terms that it develops a new e-learning artefact based on theories and models, considers customised business processes of a company, validates empirically the e-learning artefact in the context of a company with a sufficient sample size.

5.2 Suggestion. Didactic Methodical Concept, Requirements, and Expositional Presentation for Teaching an Integrated Business Process Software like SAP S/4 HANA

The suggestion relates to the proposal for the design of an e-learning artefact (Manson, 2006, p. 163). The design, previous to any development of the e-learning artefact, is based on a didactic methodical concept as well as on the requirements for building an e-learning artefact. It also considers the extended 3-2-1 expositional model of Gagné (Kerres, 2018, p. 336) that provides a structure how the didactic methodical concept as well as the requirements are embedded in this common design.

The didactic theories and models that apply for the development of the e-learning artefact should try to get the learner to learn, focus on the factors that influence learning, as well as explain the relationships among teaching objectives, teaching content and teaching methods. Upon these reasons, the design of the e-learning artefact, is based on didactic theories rather than on theories of the learning psychology. E-learning is not just an artefact for learning, it also enables the usage of information and communications technologies that are integrated with the learning processes (Seufert, Back, Häusler, 2001, p. 13). The construction of an artefact for e-learning requires knowledge related to didactic theories as well as didactic models. Additionally, it also requires a rigorous procedure to assess its quality (Meier, 2006, p. 80). The didactic methodical concept for the e-learning artefact in this project research is based on a combination of the critical and constructive didactic model from Klafki, the Berliner and Hamburger model and the APO-IT subject-didactic based on a business process orientated learning. The critical and constructive didactic model adopts didactic methodical elements from the Berliner and Hamburger didactic models. The Hamburger didactic model is an enhancement of the Berliner didactic model and includes the proof and monitoring of the learning goals from the learning-goal-orientated didactic (Riedl, 2010, p. 95). The critical and constructive didactic model is especially core as a didactic concept for the teaching planning (Riedl,

2010, p. 95). However, because an ERP system like SAP S/4 HANA orientates itself towards business process and workflows, the didactic methodical concept for the e-learning concept is combined with the subject didactic of a work process orientated teaching (APO-IT); i.e., when it comes to define the thematical structure of the learning as well as the combination of theory and praxis.

In 1985 Klafki published a perspective schema for the planning of a teaching unit with the aim of providing teachers with a planning method, also known as perspective schema, that could help them to structure their teaching (Riedl, 2010, p. 95). In this perspective schema Klafki set as a priority the definitions of the learning objectives, and later he defines the learning content, methods, and media (Plöger, 1999, p. 90). This prioritisation of learning objectives in the critical and constructive educational didactic represents a helping hand for a teacher when planning the training in terms of a subject didactic (Plöger, 1999, S. 92). Table 5.5 shows a general information of the teaching for a later planning according to Klafki's perspective schema.

Table 5.5 Overview about the general information for the teaching planning

Characteristic	Information
Date	Online course, available 24h / 7 days a week
Time	Online course, available 24h / 7 days a week
Company	Leica Geosystems AG
Trainer Manager	SAP trainer
Mentor	Operations manager of a selling unit
Learning group	Employees of the company Leica Geosystems AG
Subject	Software-as-a-Service
Theme	Create Software-as-a-Service sales order with SAP S/4 HANA
Frame objective	Employees of Leica Geosystems can process Software-as-a-Service orders in SAP S/4 HANA

The course is embedded in a e-learning artefact and available in a server in Austria. Users can access to this e-learning artefact 24 hours a day and seven days a week. The course is provided by the company Leica Geosystems AG. The corporate design of the e-learning contains the logo of the company Leica Geosystems AG, so that employees feel more identify with it. The SAP trainers are employees of the company Leica Geosystems AG, in case learners might have questions or issues related to the functionality of the e-learning artefact.

The course is created for employees of the company Leica Geosystems, independently if they work in customer care or in other areas like supply chain, service, purchasing, procurement, etc. The subject of the course is Software-as-a-Service, and the theme relates to the creation of a Software-as-a-Service sales order in SAP S/4 HANA with Fiori apps. After visiting the course, employees should be able to create and process these Software-as-a-Service sales orders in the SAP S/4 HANA system.

The Figure 5.4 shows the perspective schema of the critical and constructive didactic, upon which the didactic methodical model for the e-learning of SAP S/4 HANA is built-up. The perspective schema represents the learning conditions for the teaching planning, the context of the reasons, the thematical structure and learning objectives, the approachability and presentation possibilities, and the methodical structure. The learning conditions refer to the relevant socio-cultural aspects influencing the learning with the aim of analysing them, as well as being able to plan them an appropriate way (Riedl, 2010, p. 97). The conditions are related to the learning group, to the learners, and to the flexible or inflexible institutional given conditions (Klafki, 1985, p. 215).

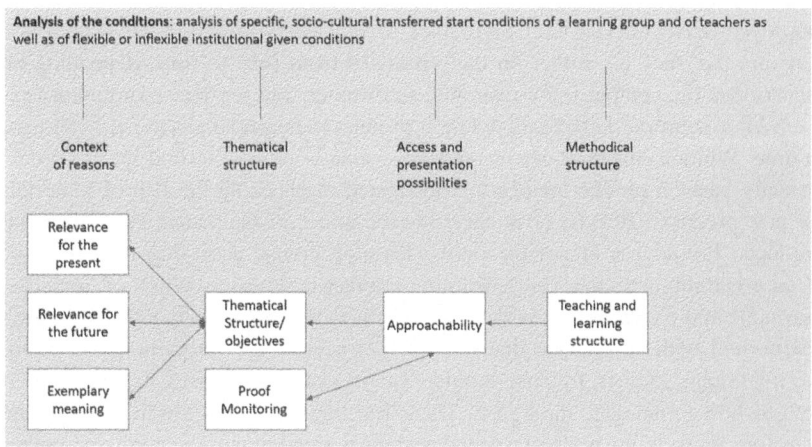

Figure 5.4 Perspective schema for the teaching planning (adapted from Klafki, 1985, p. 215)

With the analysis of the learning conditions previous to any planning, Klafki wanted to emphasise that previous knowledge and experiences of the learners play an important role when acquiring new knowledge (Plöger, 1999, S. 97). In the learning context of business processes with SAP S/4 HANA, learners are employees of the company Leica Geosystems AG working in different areas (e.g., production, technical service, customer service, procurement, purchasing, warehouse, controlling, finance, etc.) These employees come from several countries in the European Union as well as from United States of America, and Canada. While the Anglo-American countries have English as a mother tongue, the other employees working for Leica Geosystems AG in other countries have normally a high level in English. This is due to the fact that Leica Geosystems AG company's language is English. This implies that the internal communication per phone or email, as well as all internal documents and trainings are made only in English. Also, it is a requirement for every potential new employee to speak and write English on a high level. Job descriptions of the company remark also the importance of speak and write English on a high level. If English would not be a requirement for the employees, the translation of documents and trainings in several languages could cause cost inefficiencies through the whole organisation worldwide. In terms of education the employees of Leica Geosystems AG have different educational backgrounds and degrees. The educational backgrounds vary from an apprenticeship up to an university degree. Most of these employees have worked with SAP since several years and the SAP system is a software that they use either on daily basis or from time to time, depending on their role at the company. For example, a customer care service co-ordinator uses the SAP system on daily basis, while a product manager might use it from time to time. While a customer care service co-ordinator creates several service orders on daily basis, a product manager is interested in checking the Bill of Materials for new products (BOMs) or in changing the prices of the materials for different products. Employees attend the online learning course accessing the prototype of an e-learning artefact. This e-learning artefact is accessed with a personalised username and password. The e-learning artefact can be accessed at anytime, anywhere, and with any device that has internet access. Before participants utilise the e-learning artefact, trainers should explain to the employees its structure, as well as how to navigate through it. The e-learning artefact is a web solution and therefore there is no need of special technical requirements in terms of operation systems like IOS, Android, Windows, or Linux. Important though is that the learners can access to the e-learning web domain only if they are connected through company's virtual private network (VPN). This is a prerequisite for IT security purposes. Participants can, previously connected through VPN access to the e-learning artefact, also log in from any device like e.g., personal computers, laptop, tablets, or smartphones. This possibility enables participants to do the training utilising different devices.

The context of reasons addresses questions if and why the expected learning process can or should take place (Kron, 2000, p. 135). In this context Klafki (1999, p. 17) focuses on the relevance for the present, the future, as well as the exemplarity of the teaching (see Figure 5.4). The relevance for the present is related to the understanding why does it makes sense to participate in a training, and also what are the practical applications and benefits on daily basis (Klafki, 1999, p. 19). SAP S/4 HANA is an integrated process-oriented software that many employees use on daily basis to perform their work. This is applicable for several areas and in different departments. Currently at Leica Geosystems AG, nearly all employees do work more or less intensively with SAP as an ERP (Enterprise Resource Planning) system. The relevance of the training is the possibility to adapt and work with the new version of SAP called SAP S/4 HANA. As the migration from SAP ERP to SAP S/4 HANA is on track, employees need to learn it to be able to perform their work. Especially the new Fiori apps based on roles and the new graphical user interface (GUI) mean a new adaptation of their current activities with the SAP S/4 HANA system. The relevance for the future should have a direct link with the objectives and themes of interest for the learners (Klafki, 1999, p. 19). The technical migration of SAP ERP to SAP S/4 HANA or other ERP system is basically mandatory, as SAP ERP will be supported till the year 2025. The new SAP S/4 HANA is a new in-memory technology that has been and will be implemented in many companies that currently have an SAP ERP system. Employees, who today learn certain business processes with SAP S/4 HANA, will gain additional skills that they need to perform their work in the future.

With exemplarity Klafki (1997, p. 17) meant that an existing knowledge is applicable in similar areas of knowledge. Learning should help learners in their current life cycle to learn and be able to act with the learned knowledge, but learning should also open them new perspectives for the future in terms of personal development (Klafki, 1999, p. 17). The exemplary meaning refers also to a general learning topic that tries to identify which problem specifically the learning content addresses (Riedl, 2010, P.90). In this sense it also tries to answer the question what for should the theme be exemplary for. In case of SAP S/4 HANA learners will learn about the sales process and experience how to place sales orders for service software sales processes, as well as get used to the new graphical user interface with Fiori apps. This will enhance their understanding and reasoning for doing some SAP tasks in the system, and why entering some data becomes mandatory or facultative. The exemplary learning applies the way of thinking in terms of processes, and also for the utilisation of the SAP S/4 HANA software. If employees are familiar with certain business processes running under SAP S/4 HANA, then they will be able in an exemplary way to utilise it for other type of business processes. Learners will be able e.g., to apply this knowledge also when they enhance their roles and need to create other type of sales orders. Under the thematical structure understands Klafki a theme upon the objectives and the corresponding themes of learning are defined (Plöger, 1999, S. 90). For example, the theme software as a service might have, due to different objectives, also a different theme of learning for employees working in sales, in operations or in support. While the sales representatives e.g., need to explain the customer the product software-as-a-service, as well as the general terms of conditions (GTCs) of the service attached to it, employees in operations need to understand how to process a sales order in SAP S/4 HANA. Support engineers in this area need to learn how to support customers and see e.g., if a software as a service sold as a subscription is still active in the server. According to Rohs & Mattauch (p. 65, 2001), the thematical structure of APO-IT in the sense of learning objectives and themes approaches a three-step procedure for the definition of the learning objectives. This three-step procedure is related to the development of the activity roles of the learners, to the elaboration of a reference project, and to the definition of appropriate transfer projects (individual agreements on the learning objectives). The definition of the activity roles of the SAP learners at Leica Geosystems are based on the function and the qualification requirements that a person needs to fulfil. For example, for the SAP processes the following roles were defined: customer care sales co-ordinator, customer care service co-ordinator, warehouse keeper, technical service engineer, production engineer, financial manager, controller, human resources manager, product manager. Each role has its own job description that ensures consistency throughout all entities of Leica Geosystems AG. For example, the SAP role customer care sales co-ordinator is key for placing the sales orders in the area of software-as-a-service. Table 5.6 shows the corresponding job description. The elaboration of a reference project refers to a specific working process for an activity role (Rohs & Mattauch, 2001, p. 65). The process SaaS (Software-as-a-Service) is itself a sales operation process that belongs to the activity role of a customer care co-ordinator. This sales process belongs to the SAP process Order-To-Cash (see Figure 5.5). The yellow arrows represent the general Order-To-Cash (OTS) process, while the individual blue arrows below represent the needed SAP documents created in SAP. The yellow arrows in-between the blue arrows represent activities done during the creation of SAP sales process steps. The OTS process starts with an order placed by the sales representative, normally in digital form (see Figure 5.5). The sales representative processes the order and sends the order confirmation to the customer. Later, the delivery based on stock availability is prepared, the goods are issued, and the products or services are delivered. Upon confirmation of receiving the goods, the invoice is created, posted in accounting, and sent to the customer. Once the customer makes the payment, this will be posted. Figure 5.5 also shows the needed SAP process steps parallel to the OTS. The key SAP documents are the sales order, the delivery document with its integrated picking and post goods issue, the invoice, and the posting of the payment. This general OTS process is a standard process. However, this process is very general and slightly different from the software-as-a-service process (SaaS). In a SaaS process the availability of services (see Figure 5.6) is ad-hoc and the post goods issue triggers the creation of the license in the server thanks to the programming of web services. Because it is a subscription, the validity of the software is limited to a certain period. In the background, and as part of the after sales process, invoices will be automatically created until the customer terminates the subscription. However, a process itself does not contain all relevant information needed, it requires according Gadatsch (2015, p. 5) also a workflow, where other perspectives like ownership, systems and outputs are represented.

Table 5.6 Example of a job description for an SAP Customer care sales co-ordinator at Leica Geosystems

Characteristic	Information
Position title	SAP customer care sales co-ordinator
Business area	Selling unit
Department	Operations
Reporting to	Operations Manager
Location	e.g., Vienna
Role & purpose of the position	• To provide efficient and courteous customer care when required by existing and potential customers. • To provide the required office support for Leica Geosystems Limited sales personnel.
Duties & Responsibilities	• Manage the sales order process through to billing using the SAP status reports ensuring that the corporate guidelines on the method of shipment are adhered to.

(continued)

Table 5.6 (continued)

Characteristic	Information
	• Ensure that the sales orders are processed within agreed deadlines and that customers and the sales team are informed of delivery dates and any subsequent changes. • Work with Warehouse to ensure that all purchases against sales orders are confirmed for delivery within 24 hours of receipt and that the delivery dates are communicated to the customers and the sales team. • Monitor the pricing on sales orders to ensure that they are in line with the agreed price lists. • Liaise with Warehouse/Service to ensure timely dispatch of orders. • Ensure that quotations are raised by the sales team for any special pricing arrangements. • Monitor the backlog daily to ensure that confirmed delivery dates are adhered to and inform customers of any delays. • Ensure that the partner store orders are checked 3 times per day and that all orders are released promptly to minimise delays. • Assist with the prompt resolution of customer queries and credit notes to aid the cash collection process. • Ensure that all customers receive a confirmation of order and that all details are checked to minimise errors in invoicing.
Duties & Responsibilities	• Ensure that subscriptions are renewed promptly and efficiently. • Preparation any credit notes ready for authorisation then process on SAP. • Ensure that all issued information, whether written or verbal, is both accurate and authorised. • Ensure that all work is carried out in accordance with the company's quality system. • Archiving and disposal of aged records in accordance with company policy. • Other duties as may be required by the Operations Manager.

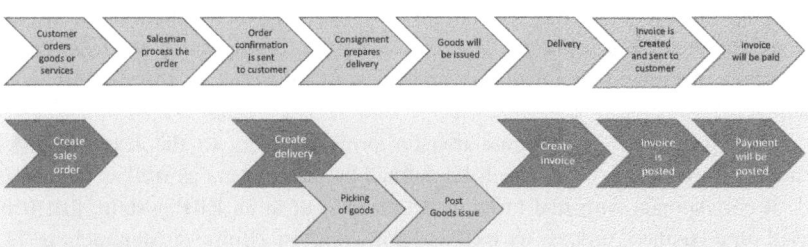

Figure 5.5 The order-to-cash process at Leica Geosystems. Analogue to SAP S/4 HANA (SAP, 2022, p. 320)

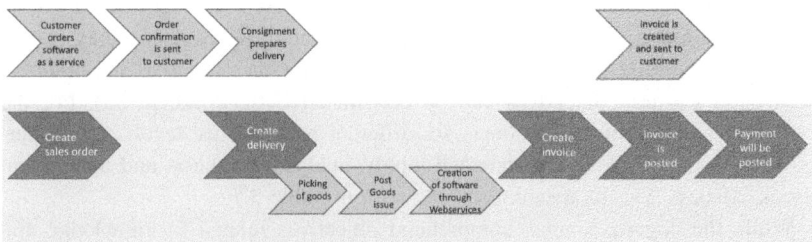

Figure 5.6 The SaaS process for Software-as-a-service. Analogue to SAP S/4 HANA (SAP, 2022, p. 320)

This workflow is based on the modelling process of EPC (event-driven process chain) already described in Section 1.2. The SaaS workflow differs from the standard order-to-cash process. The SaaS workflow shows additional perspectives of the process (e.g., output, responsibility, activity and system). Software as a product is something intangible and this has an influence in the way of processing the order in SAP as we will see in the following. In the SaaS workflow (see Figure 5.7) customers order a software. Because the sales representative uses the online e-commerce platform to order the software and see the prices in this platform, a creation of a quotation is no longer needed in SAP. After the creation of the quotation, the administration team of customer cares sales creates a sales order OR type with the Fiori app of SAP S/4 HANA. Later, an order confirmation is sent to the customer via email. The order appears in the report "orders in hand" in SAP. After this stage the administration of customer care sales processes the sales order and creates a delivery note. The goods in terms of services are then picked. After the picking of the goods in the sense of SaaS, the customer care co-ordinator posts the service. At this stage the new software license is created automatically in the software server system through web services. Later, the billing document or invoice is created with reference to the delivery note number.

The definition of appropriate transfer projects refers to the learning tasks resulting from the learning objectives defined by the teachers as well as the learners. It corresponds with real tasks to be carried out in an ERP system, that it is based on a business process as well as on a workflow (Rohs, & Mattauch, p. 72, 2001). In a transfer activity a learner should implement his or her knowledge to be able to process a business case, e.g., the process of entering a SaaS order from the creation until invoicing. The knowledge gap of learners to process this business case is determinant to define the learning objectives of the learning. When defining learning objectives, different psychological areas are considered. These areas are cognitive, affective, and psychomotor related (Hubwieser, 2000, p. 34). Benjamin Bloom developed a learning taxonomy for the cognitive and affective areas upon learning objectives can be classified (Riedl, 2010, p. 35). For the cognitive area Bloom differentiates six different performance levels. These performance levels are know, understand, apply, analyse, synthesis and assessment (Riedl, 2010, p. 35). Its meaning can be seen in Table 5.7.

While the cognitive areas comprehend objectives related to knowledge and thinking, the affective psychological area refers to the emotions and attitudes (Lehner, 2009, p. 118). Also, Bloom provided a taxonomy for the affective area and classified it in terms of observation, answer, values, assignment of values and determination of a personality through a value (Möller, 1999, p. 82). Important for the affective area is the question how learners adopt values, norms, perceptions and how these are internalised (Riedl, 2010, p. 36). The psychomotor area looks more into the motion sequence and to the motoric in general (Lehner, 2009, p. 118). It represents the technical skills needed for an action, i.e., learners can type text in a field without looking at the keyboard (Riedl, 2010, p. 34). The definition of learning objectives is more than just a way of trying to determine what the learners should achieve, but it should be measurable. This means that the operationalisation of learning objectives should consider the definition of what can be an expected and observable change in the behaviour of the learners in the cognitive, affective, and psychomotor areas (Hubwieser, 2000, p. 34). The learning objectives to process the sales order for SaaS are defined below. These objectives will consider the change in the behaviour according to the taxonomy explained above and are limited to the utilisation of an SAP S/4 HANA system as ERP software. These objectives are:

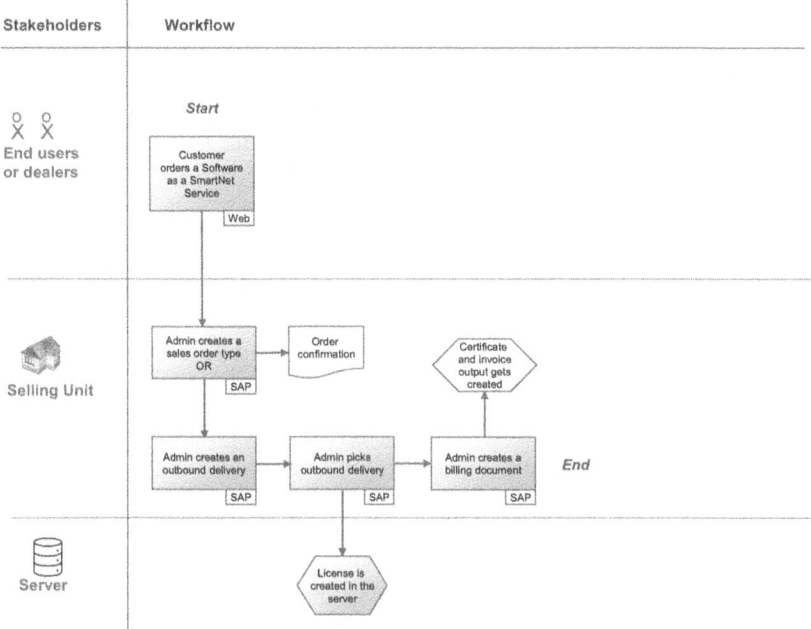

Figure 5.7   The SaaS workflow for Software-as-a-service. (Own creation)

- learners read the sales order process of SaaS (knowledge and understanding are based on performance cognitive levels),
- learners watch how the ordering process of SaaS is operationalised using SAP S/4 HANA (knowledge and understanding are based on performance cognitive levels),
- and learners process independently in SAP S/4 HANA the sales order process of a SaaS (this objective is based on performance cognitive level, as well as psychomotor area in terms that the learner is able to enter independently the data in the correct fields of the program using the computer mouse and keyboard. This objective also fulfils affective levels in terms that learners are open in their attitude and emotions to use SAP S/4 HANA to make the exercises).

The measurement if the objectives have been achieved require proof and monitoring activities. The proof and monitoring of the teaching focus on the question which competences and cognitions have been acquired, and how these should

Table 5.7 Performance levels of the learning taxonomy from Benjamin Bloom and its meaning (Riedl, 2010, p. 35)

Performance levels	Meaning
Know	If refers to knowledge where Learners learn facts and can give account of like e.g., numbers, concepts, definitions, etc.
Understand	Learners can explain issues with own words, give examples and be able to interpret problems.
Apply	It means that learners can transfer the knowledge to new situations.
Analyse	Learners can structure aspects of a problem and thus understand different structures.
Synthesis	Learners can generate new knowledge and thus also new ways of solving problems, as well as deduce hypothesis.
Assessment	Assessment enables learners to assess the value of ideas, take decisions upon founded criteria as well as consciously transfer knowledge to others.

be evaluated (Plöger, 1999, p. 97). The aim of the proof and monitoring is to identify if the learning was successful or not (Plöger, 1999, S. 97). However, a company should not only be satisfied with the fact that the learners have acquired knowledge, but also it should ensure that the evaluation is transparent and, as a continuous learning cycle, it should also utilise the findings of a training for the planning in the future (Plöger, 1999. S. 99). In the case of SAP S/4 HANA learners will place two sales orders in the system. The first sales order covers the steps from sales order creation until invoice. The second sales order goes from sales order creation until the price is changed. The trainer will check in the SAP S/4 HANA System if the learners have entered the right information in the fields to place the SAP S/4 HANA and see if they have completed the exercises.

The topic of the perspective theme looks at the way knowledge is accessed, i.e., through games, actions, but also through media like pictures, models, films. etc (Klafki, 1999, p. 29). Different learning conditions also influence which media is implemented (Klafki, 1999, p. 29). E-learning is the main media platform for the learning purposes of SAP S/4 HANA. Part of this media platform will consider the representation of the processes through Sharable Content Object Reference Model (SCORM) documents, as well as the utilisation of videos with sound. In the case of the SCORM documents, learners will read the workflow and thus gain an overview about the steps and activities that need to be done. In the case of the video with sound, learners will visualize through sequences how a sales order is created with SAP S/4 HANA. These videos will include the different steps at the time of creating the sales order until the invoice is made and posted, as well as creating a sales order and change the price. For example, there is a video for each step of the process in terms of creating sales order, create outbound delivery, pick the outbound delivery, and create invoice. The way the learner visualises the creation of an entire sales order is previous condition to create a similar sales order in the SAP S/4 HANA sandbox system.

The methodical structure refers to the learning and teaching process structure (Kron, 2000, p. 135). Klafki (1999, p. 30) is of the opinion that there is a sequence in the teaching and learning process, and that this teaching and learning process needs a methodical structure. In this sense, Klafki (1999, p. 30) states which iteration forms are needed to operationalise the teaching as well as learning process. Further, he remarks that the new research in the field of didactics could help to establish this interaction (Klafki, 1999, p. 30). Specifically, here is where the approach of the APO-IT (work process work process oriented further training in the IT) becomes relevant because the process-oriented teaching builds up also the core of the learning and the teaching process structure. A structure for a course should contain different learning phases as well as learning objectives, which are again the basement of these learning phases (Arnold, Krämer-Stürzl, & Siebert, 199, p. 94). However, the learning objectives does not only refer to the content but also the competences (professional, methodical, social, and personal) that learners need to acquire in the sense of the work orientated further training from APO-IT (Rohs, & Mattauch, p. 75, 2001). While the professional competence requires information actuality, the methodical competence focuses on building up professional action techniques through real simulation (Rohs, & Mattauch, p. 84, 2001). Social competences refer to the topic on how learners can reflect about real situations (e.g., working in team), and how personal competences can be built up based on reflexion through discussion about the learning topics (Rohs, & Mattauch, p. 84, 2001). In the structure of the course (see Table 5.8) learners first see an introduction about the process of SaaS, as well as an overview of the Fiori apps for the process, and a link to be able to access the SAP S/4 HANA system. Later, for each step of the sales order process there is a video as well as an exercise. In this way participants can watch the video on how to operationalise a step of the process first, and later make an exercise in the SAP S/4 HANA System for the corresponding step. They also can utilise existing Fiori apps that enables them to manage steps of the process. This is the case with Fiori apps related to managing sales orders or managing outbound deliveries. Through these apps participants can open e.g., a created sales order and change the price. When the learners have watched the videos with sound and made the exercises, then the course is completed.

The requirements based on scientific literature to build an e-learning artefact were described in Section 4.4. As a summary it can be said that following requirements related to the system quality, net benefits, and information quality should be considered:

- runs without any application errors and it is always available (system quality),
- is easy to use (system quality),
- helps to prepare the work packages very good (net benefits),
- helps users to learn whenever they want (net benefits),
- helps to get started very efficient with the basics of SAP (net benefits),
- provides a high content quality and the content is easy to understand (information quality),
- provides information free of contradictions (information quality)
- and the answers of the support and service administration should be quick, helpful, and understandable.

The expositional method for the e-learning artefact is based on the 3-2-1 expositional model. It considers the learning information, the learning material, the learning exercises, the communication, and co-operation, as well as the tests (see model in Section 4.5). The learning information explains the learners about the content as well as the learning objectives. The learning information in the e-learning artefact is explained through a course description. The learning materials are represented through a learning path that facilitates the learning of the process. The learning materials are structured in such a way that the e-learning artefact is able to help learners to prepare the work packages in a good way (net benefits). After each explanation of the process steps, an exercise needs to be made by the learners. The communication and co-operation take place through the chat and the forum. Learners can through the chat communicate with either the teacher or

Table 5.8   Structure of the course SaaS in an e-learning platform

Phase / Time	Learning objective (content and competences)	Material / Media
Introduction Subscription / 10 min	Learners see the process of the Subscription model to sell SaaS products, as well the Fiori apps to operationalise the process and a link to access the SAP S/4 HANA system. The process includes the steps from sales order creation until invoicing. The Fiori apps are based on this process.	SCORM document
Create a Sales Order / 10 min	Learners see how a sales order with the Fiori app "Create Sales Order" is created. Later they can do an exercise for the creation of the Sales Order in the SAP S/4 HANA test system.	Video with sound in mp4 format and text document
Create an Outbound Delivery / 10 min	Learners see how an outbound delivery with the Fiori app "Create Outbound delivery" is created. Later they can do an exercise for creating the outbound delivery in the SAP S/4 HANA test system.	Video with sound in mp4 format and text document
Pick Outbound Delivery / 5 min	Learners see how an outbound delivery with the Fiori app "Pick Outbound Delivery" is picked. Later they can do an exercise for picking the Outbound Delivery in the SAP S/4 HANA test system.	Video with sound in mp4 format and text document
Create Billing Document / 10 min	Learners see how a billing document and the posting of the billing is created with the Fiori app "Create Billing Document". Later they can do an exercise for creating the billing document and posting it in the SAP S/4 HANA test system.	Video with sound in mp4 format and text document
Manage the sales order / 5 Min	Learners understand how to manage the sales order with the Fiori app "Manage Sales Orders" that they have created in the system.	Video in mp4 format document

(continued)

Table 5.8 (continued)

Phase / Time	Learning objective (content and competences)	Material / Media
Create a new sales order and change the price / 10 min	Learners see how a new sales order with the Fiori app "Create Sales Order" is created and how to change the price with the Fiori app "Manage Sales Order". Later they can do an exercise for sales order as well as changing the price in the SAP S/4 HANA test system.	Video with sound in mp4 format and text document
Manage the outbound delivery / 5 Min	Learners understand how to manage outbound deliveries with the Fiori app "Manage Outbound Deliveries" that they have created in the system.	Video with sound in mp4 format

with other students. They also can make through the forum statements or place open questions to other learners. Instead of the tests, participants who complete the exercise receive a certificate, as stated by Kerres (2018, p. 338).

5.3   Development

The 3-2-1 expositional model of Gagné (Kerres, 2018, p. 336) provides a structure how the didactic methodical concept as well as the requirements are embedded in a common design. Based on this common design, the development represents the build-up of the e-learning artefact (Manson, 2006, p. 163). After logging into the course, the structure begins with a welcome page about the course under the tab "Homepage" (see Figure 5.8). In this "Homepage" users see a first general objective (learning information) as well as a first point of contact (communication) before logging into the e-learning artefact. Users can see also on the left-hand side the potential integration of their e-mail accounts including where invitations for the courses are sent to, general personal data like username and password, and the possibility of editing certain data of their user profiles like e.g., password to log into the system. The potential integration of the e-mail account gives the users the possibility of directly writing an email to the system administrator and this make the e-learning artefact easy to use (system quality). In case that the system administrator is not available due e.g., to holidays, his or her email will be forward to a deputy, who can then help the learner. The personalisation of the system administrator is important to ensure that the answers will be helpful and quick (service quality). The reason for this is that the more the name of the supporter is known, the higher the commitment and responsibility towards understandable answers.

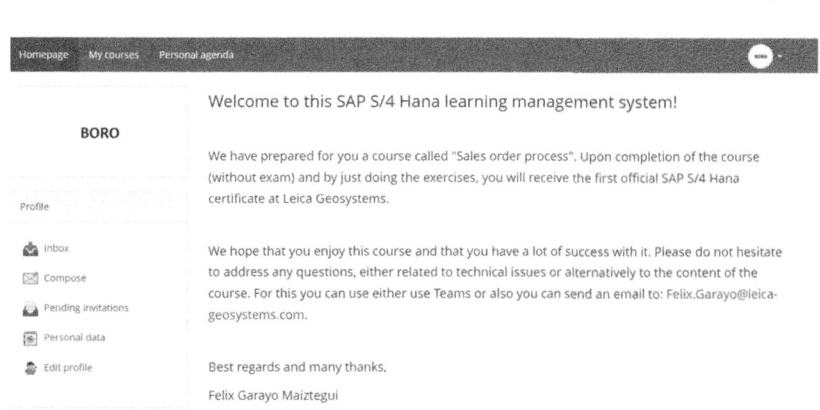

Figure 5.8 Welcome page of the e-learning artefact. (Own creation)

Later, under the tab "My courses", participants can see the course that needs to be completed. In the context of this dissertation the course is called "Sales Order Process" (see Figure 5.9).

In the future it is expected that users might have several courses based on their different roles. By clicking onto the link "Sales Order Process", learners jump into the content of the course divided in four areas with four apps. These apps are covered by the topics course description, learning path, forums, and chat (see Figure 5.10).

The description (see Figure 5.11) provides the objectives of the learning as well as the general content of the course (learning information). The SAP content structure of the course (learning materials) is based on the "learning path", while the communication and co-operation is provided the "forum" and the "chat". The "learning path" (see Figure 5.12) represents the structure of the sales order process with the corresponding workflow steps, as well as exercises related to these steps. By clicking onto the "learning path", the e-learning artefact displays the number of lessons, as well as the progress. With the "learning path" the users are guided through the content. The lessons units are numerated from U0 (unit

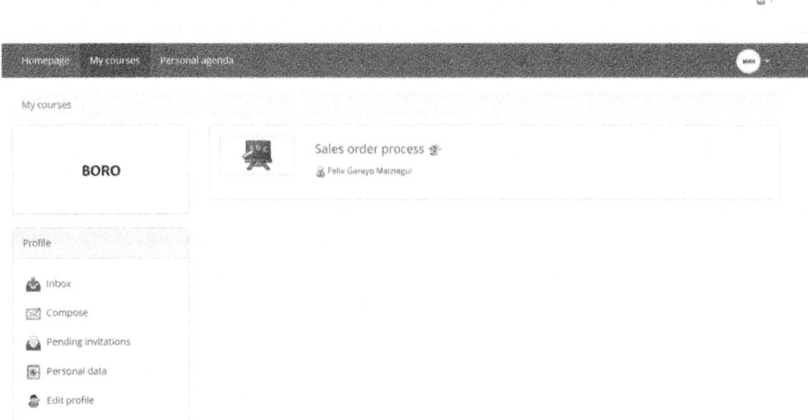

Figure 5.9   Sales order process of My courses. (Own creation)

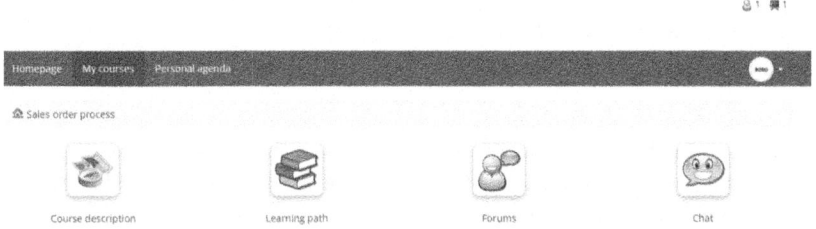

Figure 5.10   Overview of course for the Sales order process. (Own creation)

0) until U05 and include an exercise at the end of each unit. Unit 0 explains the learners the sales order workflow with its steps for software subscriptions and gives then an overview of these steps in an overview with Fiori apps. Also, it includes a unit explaining them how to log into SAP S/4 HANA (see Figure 5.13). By clicking into the link, the user of the e-learning artefact jumps directly into the SAP S/4 HANA productive system (see Figure 5.14).

The reason behind this structure is that learners need to understand, independently of the ERP system, how the workflow is built-up. In that way they can follow the steps with the Fiori apps later. Thus, employees will not reduce their activities to just open Fiori apps and enter data in different fields, but they will

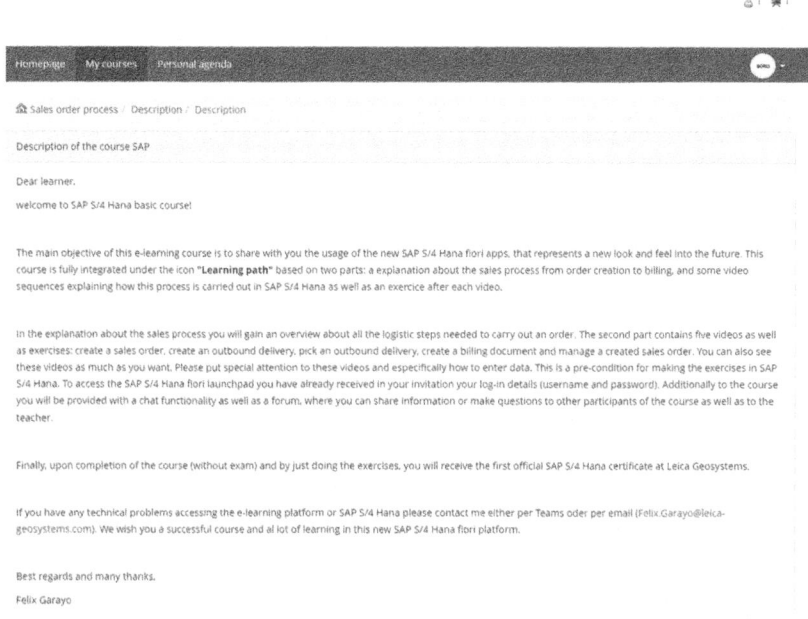

Figure 5.11 Description of the sales order process course. (Own creation)

understand that they operatively are completing a step of a sales process. This abstraction of knowledge is crucial for the relevance of the teaching (see didactic methodical concept in Section 5.2). Once employees have understood the sales order workflow as part of the process and have started to get familiar with the structure of the Fiori apps, they can log into the SAP S/4 HANA system. The following units of the "learning path" corresponds with the different steps of the process.

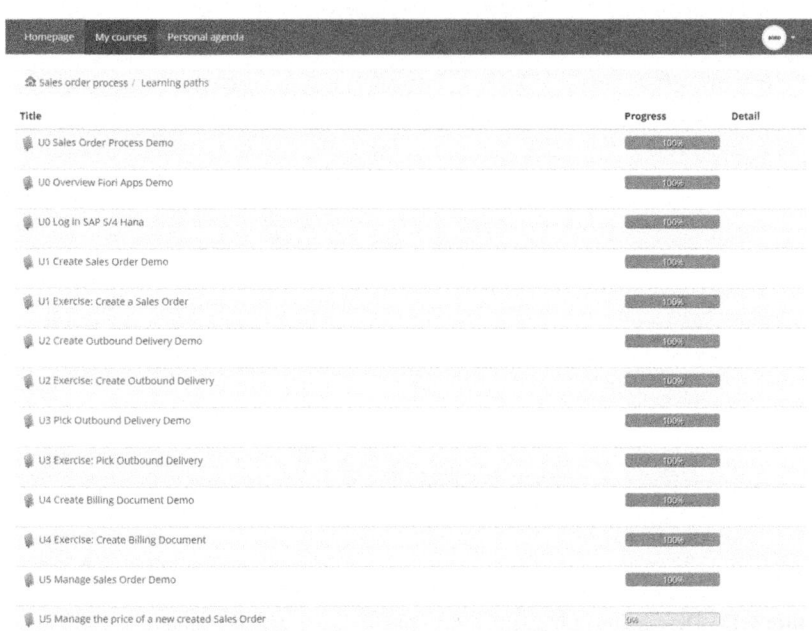

Figure 5.12   Structure of the sales order process course. (Own creation)

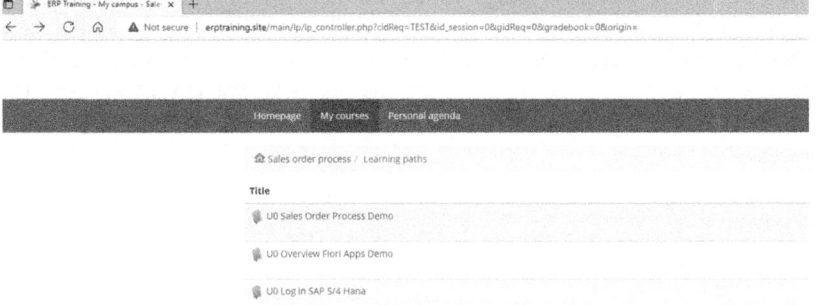

Figure 5.13   Link for log in SAP S/4 HANA

Figure 5.14   SAP S/4 HANA System

Each unit contains the same structure. First, there is a demonstration through a video with sound about the process step to be operationalised in SAP S/4 HANA. Second users can process this step themselves in SAP S/4 HANA. Users can watch the videos as much as they want. This helps them to prepare the work packages in a good manner (net benefits). With this structure participants can get started very efficient the basics of SAP (net benefits), follow the structure as well as also see the progress of the units. This is quite important in case participants need to interrupt the learning due to major reasons. Especially during work sometimes there could be urgent, important, and unexpected tasks, that might cause this interruption. The e-learning artefact allow users to learn whenever they want (net benefits). There are also in the course two apps related to the communication and co-operation. These are the apps "forum" and "chat". The "forum" app allows learners to give feedback or post opinions about the e-learning artefact or the content of it (see Figure 5.15). Finally, under "chat" app employees can interact with each other and send messages either to the tutor or to other colleagues. This is crucial, when participants are facing problems (e.g., if the videos would load too slow, or users would not be able to log into SAP S/4 HANA) and need technical support from the tutor (service quality). In this way answers of the support and service administration can be quick (service quality).

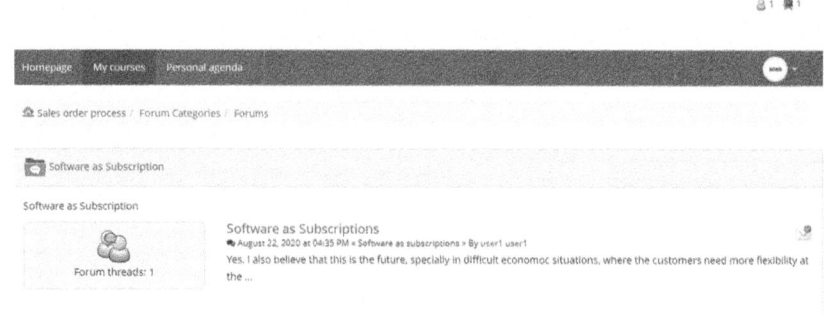

Figure 5.15 Forum of Sales order process course. (Own creation)

The exercises done in SAP S/4 HANA should help participants learn how the SaaS order is processed. Later the exercises will be evaluated directly in SAP S/4 HANA system to see if the knowledge has been transferred, and also see how good the performance was. Participants who complete the exercises got a certificate of participation signed up by the manager of the SAP trainer team as well as the Process Manager for SaaS sales orders. The evaluation of the knowledge and the issue of official certificates (see Figure 5.16) by Leica Geosystems increases the recognition of the e-learning artefact and the future value (see relevance for the future in Section 5.2) for making the course (Kerres, 2018, p. 338).

The e-learning artefact itself was programmed using the software Chamilo, an open-source e-learning platform and content management system based on Hypertext Preprocessor PHP (Wikipedia, 2022, Chamilo, Para. 1). PHP is a script programming language used mostly to program websites or web applications (Wikipedia, 2022, PHP, para. 1). Chamilo can be installed in a private server or in the cloud. For this research project Chamilo was installed in a server called ftp.erptraining.site. The users received the link http://www.erptraining.site to be able to access to the e-learning solution. The software SAP EnableNow was used to create the SCORM files as well as the videos. SAP EnableNow is a software of the company SAP that helps to record the content of software and create software animations (see Figure 5.17).

Figure 5.16   Example of a certificate. (Own creation)

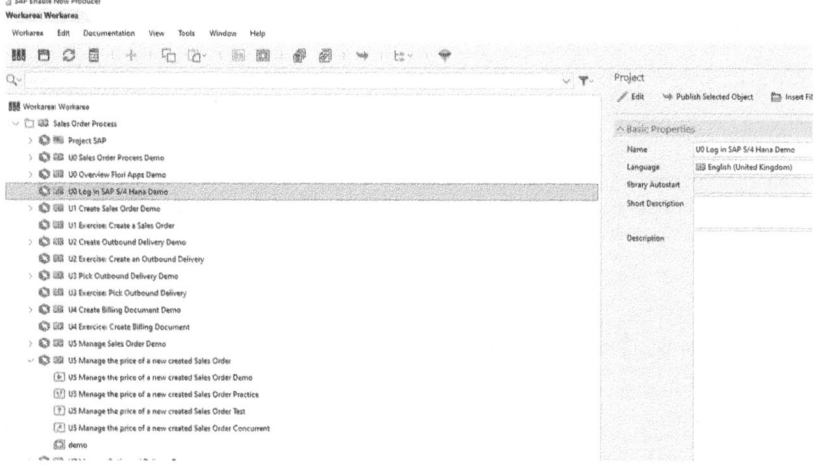

Figure 5.17   Program EnableNow producer. (Own creation)

SAP EnableNow software creates automatically sequences in a recording that can later also be seen in a video format (see Figure 5.18). The software also gives the possibility to save these sequences in a Word document or PowerPoint format automatically. This is ideal as training documentation, because it can be further kept and be provided in different way to the users. One of the key assets of SAP EnableNow is that a record session can be adapted and changed at a specific time of a sequence. Later, the adapted sequence can be recorded automatically again in a video, in a SCORM format, or in PowerPoint or Word document format. This feature of changing existing process documentation is crucial because business processes are changed and adapted continuously according to customers and market needs. The possibility of adapting the documentation at a specific time of a sequence and the corresponding automatic saving in a video is a very efficient way of adapting documentation. This reduces the time of changing documentation considerably, and at the same time reduces also costs in a company enormously. Both videos and files are stable in terms that they run without any application errors (system quality) and are also easy to use (system quality).

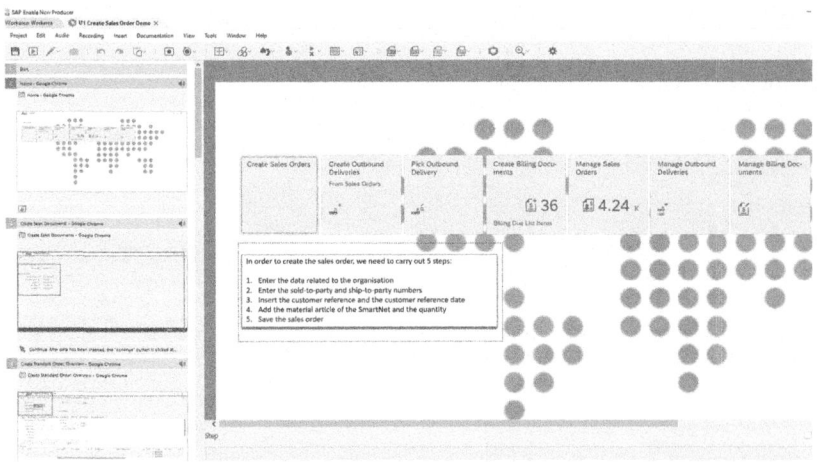

Figure 5.18 Program SAP EnableNow producer with process step "create sales order". (Own creation)

5.4 Empirical Design and Evaluation Concept

Once the e-learning artefact has been developed and implemented, it needs to be evaluated (Manson, 2006, p. 163). Previous to the development of the artefact hypothesis researchers built-up hypotheses and in positivists research these hypotheses are falsified or validated. (Manson, 2006, p. 163). Also, as stated in Section 2.1, rigorous inquiry starts with an indeterminate situation that causes the researchers to doubt something, from this doubt they formulate a question, further they create potential solutions, they test these solutions via experimentation (positivist approach) and gather empirical evidence (Korte, & Mercurio, 2010, p. 64). This chapter describes the hypotheses, variables and indicators, as well as the experimental research design, and the evaluation of the artefact.

5.4.1 Hypothesis, Variables and Indicators

There are different approaches for the validation of an artefact like e.g., observational, analytical, experimental, testing, and descriptive (Hevner, March, Park, Ram, 2004, p. 86). For the validation of the e-learning SAP S/4 HANA, the experimental approach is utilised, because this is appropriate to measure qualities (e.g., usability) in a controlled environment (Hevner et al, 2004, p. 86). The empirical design is based on hypothesis testing with a randomised field experiment. A hypothesis is a suggested answer to a defined problem (Tuckman, & Harper, 2012, p. 85) and should have the following characteristics. It should be testable empirically (Atteslander, 2010, p. 45), clear and unambiguous, and it should conjecture the relation between two or more variables (Tuckman, & Harper, 2012, p. 86). In this sense, and according to the research question in Section 1.3, there will be two types of hypotheses to be tested. These are effectiveness and efficiency related. Effectiveness is understood as missing tasks, while doing the exercises, in terms of dropouts. This means employees who could not complete the exercises. Also, there is an analysis of the dropouts to identify potential causes, e.g., SAP knowledge or time needed to carry out the exercises. For the effectiveness following hypothesis has been defined:

- H0: The implementation of an e-learning artefact does not reduce the dropout quantity in the usage of SAP S/4 HANA Fiori apps.
- H1: The implementation of an e-learning artefact reduces the dropout quantity in the usage of SAP S/4 HANA Fiori apps.

Efficiency is measured in terms of time that an employee needs to complete an exercise. For the efficiency the following hypotheses was set:

- H0: The implementation of an e-learning artefact does not positively influence efficiency in the usage of SAP S/4 HANA Fiori apps.
- H1: The implementation of an e-learning artefact positively influences efficiency of SAP S/4 HANA Fiori apps.

According to Stenberg & Stenberg (2012, p. 188) there are two types of knowledge: declarative and procedural knowledge. Declarative knowledge is knowing something that is a fact, e.g., first name of a person, and procedural knowledge is a skill to do something in terms of knowhow (Matošková, 2016, p. 7). While the task required for measuring memory in the case declarative knowledge implies to recall a fact, measuring memory in terms of procedural knowledge requires that the learner must remember learned skills and automated behaviours more than facts (Stenberg, & Stenberg 2012, p. 188). In this research project procedural knowledge as well as declarative knowledge will be measured. For the procedural knowledge the learners will need to remember the steps needed to e.g., create a sales order, create outbound delivery, pick outbound delivery, and create billing document. For the declarative knowledge the learners will need to remember the new price of a SmartNet product. According to Tuckman and Harper (2012, p. 67) the independent variable is a stimulus variable within an environment and is a factor selected by the researcher to observe a relationship to an observed phenomenon. In this research project the independent variable is the treatment. Also, for this research project it is aimed to measure knowledge (term). Dependent variable for knowledge is attested performance. This attested performance will be measured with indicators. Indicators are variables that are directly observable (Atteslander, 2010, p. 48) in the SAP S/4 HANA System. These variables are also dichotomic in terms of exist or non-exist. For example, if the sales order is saved, then it exists. Also, if the price has been changed in the field, then it exists. The indicators are saved sales order, saved outbound delivery, saved pick outbound delivery, and saved billing document. These indicators are manifest variables in terms of metrical variables that can be observed directly in the SAP S/4 HANA System (Atteslander, 2010, p. 48). It will be based also in terms of existing or not existing (see Table 5.9), as well as in terms of duration for creating the different documents (see Table 5.10).

Table 5.9 Manifest variables for dropouts. (Own creation)

Manifest variables	Result	Points for existing or document saved	Points for inexistent or document not saved
Sales organisation	Sales organisation entry	1	0
Distribution channel	Distribution channel entry	1	0
Division	Division entry	1	0
Sales Office	Sales office entry	1	0
Sales group	Sales group entry	1	0
Sold-To-Party	Sold-to-party entry	1	0
Ship-To-Party	Ship-to-party entry	1	0
Customer Reference	Customer Reference entry	1	0
Customer Reference date	Customer Reference date entry	1	0
Sales order	Saved sales order	1	0
Outbound delivery	Saved outbound delivery	1	0
Pick outbound delivery	Saved pick outbound delivery	1	0
Billing document.	Saved billing document	1	0
Post billing document	Posted billing document	1	0
Create a new sales order	Saved sales order	1	0
Change price of a sales order	Price of sales order changed is not 150 USD	1	0
Correct price change	Price change from 150 USD to 300 USD	1	0
Drop-out process	Incompletion of sales orders	1	0
Sum		18	0

Table 5.10 Manifest variables for time duration. (Own creation)

Manifest variables	Result	Time Duration
Sales order	Time in creating a sales order	00:00:00
Outbound delivery	Time in creating an outbound delivery	00:00:00
Pick outbound delivery	Time in creating in picking an outbound delivery	00:00:00
Billing document	Time in creating a billing document	00:00:00
Create a new sales order and change the price	Time in creating a sales order and changing the price	00:00:00
Sum		00:00:00

5.4.2   Experimental Research Design

The field experiment is based in a comparison between two operational programs: an online instruction and the e-learning artefact. An online instruction,
also called group instruction, is when an instructor delivers verbally a training
to several learners at the same time (Tuckman, & Harper, 2012, p. 108). On
the opposite side, an individualised instruction is a software designed instruction
by a researcher, where students work at their own pace (Tuckman, & Harper,
2012, p. 108). The instruction through an e-learning artefact is called individualised instruction because learners can work with a device (e.g., computer, tablet,
smartphone) by themselves at their own pace (Tuckman, & Harper, 2012, p. 108).
The design for the evaluation of the artefact is based on a posttest-only control
group design (Tuckman, & Harper, 2012, p. 152). This design represents the
comparison of two groups, one that experience the treatment with an e-learning
artefact (treatment group), and one that does not (control group). The experiment as posttest-only control group design and its data analysis focus on the
comparison between the mean of O1 and O2 (see Figure 5.19). The field experiment applies for T (treatment) group as well as C (control) group. Historically
the company Leica Geosystems has been doing the SAP trainings online with a
word documentation. In this research project a new e-learning artefact has been
developed. While the C group is represented through the employees, who kept
receiving the training online with a word document, the T group made the training
with the e-learning artefact.

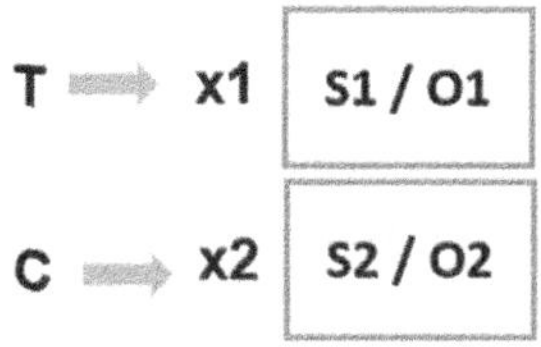

Figure 5.19
Posttest-control group
design (Tuckman, &
Harper, 2012, p. 152)

T $\implies$ x1 | S1 / O1

C $\implies$ x2 | S2 / O2

The current teaching documentation of the control group is based on the behaviourism learning theory. In the behaviourism learning theory, the view of learning is a passive way of knowledge absorption by the learner, where repetition is promoted and the teacher transfer the correct response (Radianti, Majchrzak, Fromm, Wohlgenannt, 2020, p. 3). In the case of the control group, employees receive in a word document the learning content, upon which they can do the exercises. The e-learning artefact, as explained in chapter 4 referred to the DSR knowledge base, is based on a didactic methodical concept, in the ISSM requirements, and the 3-2-1 expositional model. For both T and C groups, based on samples S1 and S2, an objective and subjective test was carried out. The samples S1 and S2 are based on a sample population of 440 employees from different Organisations of the company Leica Geosystems in Europe as well as US & Canada which have experience working with SAP ERP, however, they are supposed not to have experience either creating sales orders and/or having ever used fiori apps from SAP S/4 HANA. This is known due to the created SAP documents that we can see in the productive SAP live system from all SAP users (objective test). However, it is unknown if someone of this sample, outside the company Leica Geosystems, could have worked with the SAP sales order process or made externally a course with the SAP S/4 HANA fiori apps. To investigate this, avoid potential bias, as well as try to homogenise the sample as much as possible, a small questionnaire (subjective test) was sent to the potential learners. This questionnaire contains the following questions and was carried out online per internal email (see Appendix C in additional electronic material). The questionnaire was answered by all 440 employees in the period for 2 months, from beginning of February until March 2021. Although all employees answer the questionnaires, some of them had to be chased two or three times. The sample size of 325 employees is the result of the sample population minus the employees that have experience with the sales process or have worked with SAP S/4 HANA Fiori apps (also subjectively checked). The sample size follows the principle that the bigger the sample size, the better the estimation (Diekmann,1998, p. 189). The validation of the artefact was carried out among the employees (learners) of

Leica Geosystems. An employee is an adult who currently works more than one year at the company. In a year an employee is able to carry out SAP standard process activities as good as a more experienced employee. Once the questionnaires were collected, both groups T and C were observed through the analysis of the data. The employees were randomly entered in an excel list and uploaded in the SPSS list and clusters of observations were created (Tuckman, & Harper, 2012, p. 290). This gave a possible homogeneous starting situation of the sample size for S1 and S2. The random criteria selected is based on the answer of the questionnaire in terms of the principle "first answers, first entered in the excel list". With this approach variables can be neutralised (e.g., gender), this means technically speaking, that there are third variables which do not correlate with the experimental factor due to the randomisation (Tuckman, & Harper, 2012, p. 297). In this case issues due to the spurious correlation can also be avoided (Tuckman, & Harper, 2012, p. 297).

The clusters were built up according to the hierarchical analysis made with SPSS based on binary numbers relate to different working areas, e.g., customer care service, warehouse, etc. Based on these clusters, the groups were created and assigned in the control and treatment groups. For this purpose, 4 to 8 clusters were originally created. The result shows that 4 cluster represent for the experiment the best results in order to classify them in control and treatment groups (see Table 5.11).

Table 5.11 Hierarchical Clusters with SPSS. (Own creation)

4 Cluster	44	90	96	95				
5 Cluster	44	90	59	95	37			
6 Cluster	44	29	59	95	37	61		
7 Cluster	44	29	59	95	37	43	18	
8 Cluster	27	29	59	17	95	37	43	18

The first half of each cluster was selected for the control group and the second half of each cluster was selected for the treatment group. Additionally, it is the aim in this research project to avoid the so-called teaching effect during the SAP training. The teaching effect is given, when the researcher expects a result and unconsciously influence the employees during the online instruction either verbally, through mimic or body language (Brosius, Haas, Koschel, 2012, p. 212). To avoid this teaching effect, three additional trainers on top of the researcher of the company Leica Geosystems carried out an online instruction for the control group. These three trainers were two trainers from the Leica SAP organisation in Spain, and a trainer from the Leica SAP organisation in India. Each of the trainers did one fourth of the instruction. Further, all the online instructions were carried out based on a standardised predefined text for the treatment and the control group to ensure that the trainers explain exactly the same documents and use the same wording. Thus, during the training, the trainers read through the text. After the training, either through online or through the e-learning artefact, employees had to carry out some exercises in SAP S/4 HANA. In this sense time was also measured for every step of the process (i.e., create sales order, create outbound delivery, pick outbound delivery, create invoice, and post it, create a new sales order and change the price).

The SAP S/4 HANA system was a sandbox system where users could log in and make the exercises. For this SAP S/4 HANA system only 10 licenses were available. This means that only 9 participants could at the same time participate in the training (one license was for the trainer). Participants had one week time to complete the training including the exercises. Based on the number of the sample size with 325 employees, the experiment took 10 months, from March 2021 until January 2022. Of 325 employees, 254 participated in the experiment (129 employees for the treatment group, and 125 employees for the control group). The employees that participate in this experiment were from the selling units of the company Leica Geosystems AG in Austria, Belgium, Canada, Denmark, Italy, Finland, France, Germany, Netherlands, Norway, Poland, Portugal, Spain, Sweden, Switzerland, United Kingdom, and United States of America. These employees worked in several departments like controlling, finance, human resources, technical service, procurement, production, product management, purchasing and warehouse. The training was carried out only in English language.

Evaluation of the Artefact

6

This chapter aims to share the results of the statistics for the experiment done with a treatment and a control group at the company Leica Geosystems AG. Basically, statistics can be classified in descriptive and correlative statistics. While the descriptive statistics represents the measurement of location scales and dispersion measures for individual variables, the correlative statistics are used for deductively obtained (statistical) hypotheses with several variables (Brosius, F., 2013, p. 379). For these statistics both types were considered. The results of this experiment are based on two types: on efficiency and on effectiveness results. Efficiency is measured in terms of time that an employee needs to complete an exercise. In this experiment employees of the company Leica Geosystems had to complete two major exercises. The first exercise contained four steps. These steps were to create a sales order, an outbound delivery, pick the outbound delivery and create a billing request. In a second exercise the employees had to create a new sales order and change the price. While doing these exercises, the time was measured in both control and treatment group. For the efficiency the following hypotheses was set:

- H0: The implementation of a e-learning artefact does not positively influence efficiency in the usage of SAP S/4 HANA Fiori apps.
- H1: The implementation of a e-learning artefact positively influences efficiency of SAP S/4 HANA Fiori apps.

Effectiveness is understood as a result of missing tasks in terms of dropouts, while doing the exercises. This means, employees who could not complete both exercises mentioned above. Also, there is an analysis of the dropouts to identify potential causes, e.g., SAP knowledge or time needed to carry out the exercises. For the effectiveness the following hypothesis has been defined:

© The Author(s), under exclusive license to Springer Fachmedien Wiesbaden GmbH, part of Springer Nature 2023
F. Garayo Maiztegui, *Design and Evaluation of an E-Learning Artefact for the Implementation of SAP S/4 Hana®*, Gabler Theses,
https://doi.org/10.1007/978-3-658-40731-5_6

- H0: The implementation of a e-learning artefact does not reduce the dropout quantity in the usage of SAP S/4 HANA Fiori apps.
- H1: The implementation of a e-learning artefact reduces the dropout quantity in the usage of SAP S/4 HANA Fiori apps.

6.1   Efficiency Related Results

To see if the alternative H1 hypothesis is validated, SPSS statistics were used. Because it is an experiment with two groups, the first thought was to make a t-test. A t-test assesses whether the means of two groups differ statistically from each other (Stockemer, D., 2019. p. 101). To carry out a t-test several conditions must be met. According to Stockemer (2019, p. 101) the conditions are: the dependent variable should be continuous, all observations should be independent, there should not be many significant outliers, the dependent variable should be normally distributed and the variances between groups should be similar. The dependent variables values lay between 0 and 1, which means that they are continuous. Also, the observations are individual and independent, meaning that there is no linkage or direct influence of the values in the observations within the groups. However, whether the existence of outliers or if the dependent variables are normally distributed is something to be checked.

The Q-Q Plot of point in Figure 6.1 shows that the variables related to the time in terms of creating a sales order (salesorderseconds) create an outbound delivery (outboundseconds), pick an outbound delivery (pickedseconds), create a billing (billingseconds), as well as create a new sales order and change the price (newsalesorderseconds), are not normally distributed, and have several significant outliers. Also, normal distribution tests based on Kolmogorov-Smirnoff-test and the Shapiro-Wilk-test were carried out to check the normal distribution of the variables. The significance values in the Table 6.1 show values smaller than .001, which means that the variables are not normally distributed (Brosius, F., 2013, p. 405).

The Q-Q normal distribution tests as well as the Kolmogorov-Smirnov and the Shapiro-Wilk Test show that variables for efficiency are not normal distributed and show a big difference. In this case a Mann-Whitney-U test as well as Kruskal-Wallis-H test were carried out, to see if the groups in terms of time efficiency differentiate themselves (Bühl, A., 2016, p. 360). The less needed time in terms seconds for the treatment group is especially remarkable and significant (see Tables 6.2 and 6.4) when creating the outbound delivery (<.001), picking the outbound delivery (<.001) and creating the billing document (<.001). All these

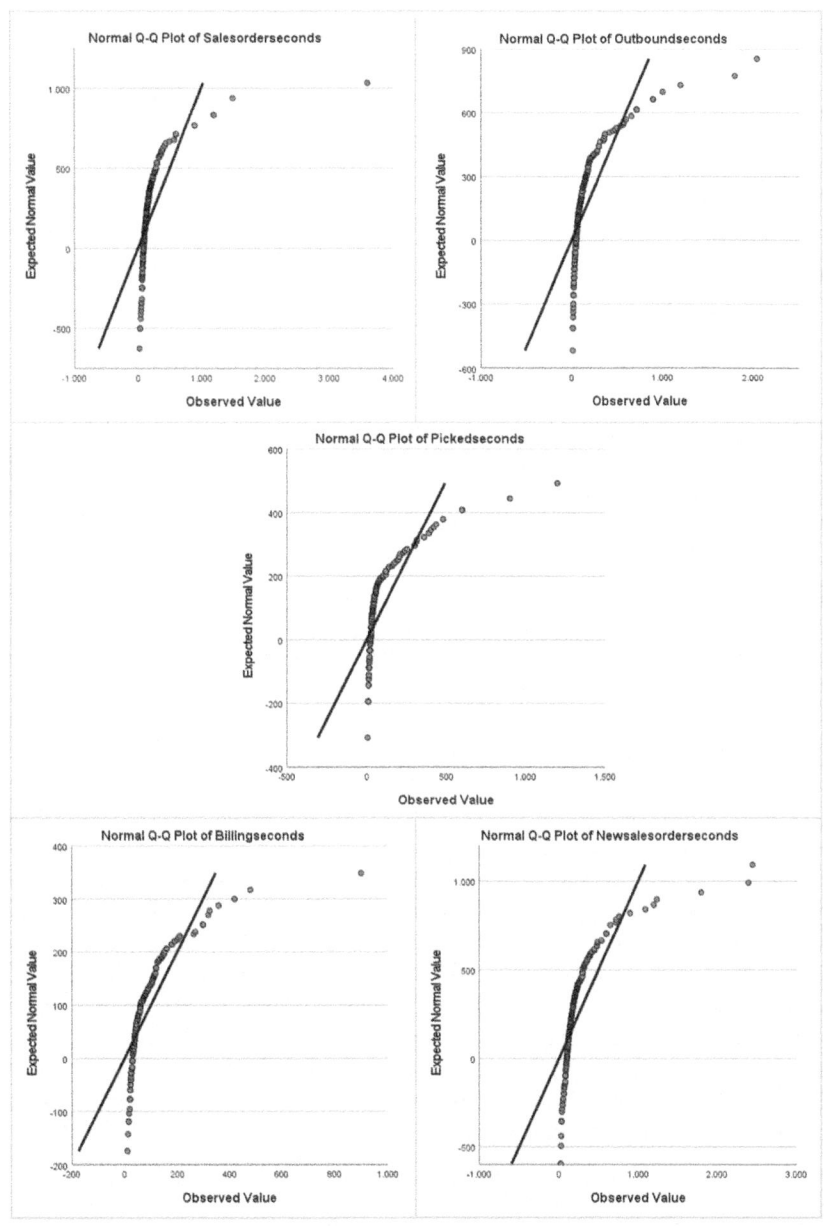

Figure 6.1   Normal Q-Q-Points. (Own creation)

Table 6.1   Test of normality. Kolmogorov-Smirnow and Shapiro-Wilk tests. (Own creation)

Tests of Normality

	Kolmogorov-Smirnov[a]			Shapiro-Wilk		
	Statistic	df	Sig.	Statistic	df	Sig.
Salesorderseconds	.243	188	.000	.572	188	.000
Outboundseconds	.263	188	.000	.512	188	.000
Pickedseconds	.290	188	.000	.626	188	.000
Billingseconds	.224	188	.000	.603	188	.000
Newsalesorderseconds	.222	188	.000	.553	188	.000

a. Lilliefors Significance Correction

tasks were part of the first exercise. Table 6.2 shows also the value Wilcoxon-W with the sum or the ranges (Brosius, 2013, p. 879) as well as the value Z significant differences between the two groups. The lower mean rank values in the Table 6.3 and 6.5 show that treatment group in all the exercises needed less seconds than the control group in terms of creating the sales orders, the outbound deliveries, picking the outbound deliveries, creating the billing requests, as well as creating new sales orders and changing the price. Especially the difference of mean ranks values between the treatment and the control groups are bigger when creating the outbound deliveries, pick the outbound deliveries and creating the billing request. The second exercise of creating a new sales order and changing the price does not show a high significance value (0.06), however a positive trend. The question that addresses from this outcome is why in these three tasks of the exercise the difference is bigger than creating the sales orders. The creation of a sales order is always the first step of the exercise on both exercises. As employees go through the exercise, the treatment group might find it easier to perform the subsequent exercises quicker with the e-learning artefact that contains videos in comparison to those employees who have to refresh their knowledge with the documentation and the notes in the control group. The decay and cognitive remembering of information over time (Weiten, W., 2008, p. 208) could explain why the treatment group with the e-learning artefact could perform the exercises quicker as they went through the exercises. The results of the Mann-Whitney-U test and Kruskal-Wallis-H test confirm the hypothesis that the implementation of an e-learning artefact positively influences efficiency utilising SAP S/4 HANA Fiori apps.

Table 6.2 Efficiency, time, Mann-Whitney-U tests. (Own creation)

Test Statistics[a]

	Salesorder-seconds	Outbound-seconds	Pickedsec-onds	Billingsec-onds	Newsalesorder-seconds
Mann-Whitney U	7085,500	4649,000	3728,000	3609,500	4806,500
Wilcoxon W	15213,500	11789,000	10749,000	10395,500	11247,500
Z	−1,059	−3,786	−3,997	−3,689	−1,879
Asymp. Sig. (2-tailed)	,290	<,001	<,001	<,001	,060

a. Grouping Variable: Group

Table 6.3 Efficiency, time, Mann-Whitney-U tests. Ranks. (Own creation)

Ranks

	Group	N	Mean Rank	Sum of Ranks
Salesorderseconds	Treatment	127	119,79	15213,50
	Control	121	129,44	15662,50
	Total	248		
Outboundseconds	Treatment	119	99,07	11789,00
	Control	110	132,24	14546,00
	Total	229		
Pickedseconds	Treatment	118	91,09	10749,00
	Control	93	124,91	11617,00
	Total	211		
Billingseconds	Treatment	116	89,62	10395,50
	Control	89	120,44	10719,50
	Total	205		
Newsalesorderseconds	Treatment	113	99,54	11247,50
	Control	100	115,44	11543,50
	Total	213		

Table 6.4   Efficiency, time, Kruskal-Wallis-H tests. (Own creation)

Test Statistics[a,b]

	Salesorder-seconds	Outbound-seconds	Pickedsec-onds	Billingsec-onds	Newsalesor-derseconds
Kruskal-Wallis H	1,122	14,332	15,976	13,609	3,532
Df	1	1	1	1	1
Asymp. Sig.	,290	<,001	<,001	<,001	,060

a. Kruskal Wallis Test
b. Grouping Variable: Group

Table 6.5   Efficiency, time, Kruskal-Wallis-H tests. Ranks. (Own creation)

Ranks

	Group	N	Mean Rank
Salesorderseconds	Treatment	127	119,79
	Control	121	129,44
	Total	248	
Outboundseconds	Treatment	119	99,07
	Control	110	132,24
	Total	229	
Pickedseconds	Treatment	118	91,09
	Control	93	124,91
	Total	211	
Billingseconds	Treatment	116	89,62
	Control	89	120,44
	Total	205	
Newsalesorderseconds	Treatment	113	99,54
	Control	100	115,44
	Total	213	

6.2   Effectiveness Related Results

The effectiveness related results were analysed in terms of dropouts, this means employees who did not complete the exercises. As a first step the intention was to utilise the Chi-Square test to see if the treatment group has less dropouts that the control group. The Chi-Square test requires some conditions: the sample is random, it should have at least 6 fields and the expected frequency should be bigger than 5 (Brosius, F., 2014, p. 237). Although the sample is random, it can be said that the maximal fields achieved are four (treatment and control group, as well as a dropout or non-dropout). For this reason, and according to the literature, the McNemar test (Brosius, F., 2013, p. 444) was utilised. The McNemar test is based on the Chi-Square test and it is appropriate for two binary variables. Additionally, the Phi as well as the V-Cramer tests were utilised to see how strong this relationship between the variables is (Brosius, F., 2013, p. 434). The McNemar test shows a high significance (p < .001) in terms that the treatment group had significantly less dropouts in comparison to the control group (see Table 6.6).

Table 6.6 Effectiveness, dropout, McNemar test. (Own creation)

Chi-Square Tests

	Value	Exact Sig. (2-sided)
McNemar Test		<,001.[a]
N of Valid Cases	254	

a. Both variables must have identical values of categories.

More specifically Table 6.7 shows that the percentage of dropouts for the treatment group sits by 19,4%, while the control group counts for 39,2%. Another aspect is to identify how strong this relationship is. For this purpose, the Phi Test was implemented. While the Phi value is adequate for binary variables, V-Cramer table is adequate for binary as well as other type of variables (Brosius, F., 2013, p. 433). According to Brosius (2013, p. 434), Phi and V-Cramer values of 1 would show a very strong relationship, however this value is hardly achieved. A value of 0.218 shows a moderate relationship (see Table 6.8).

Table 6.7 Effectiveness, dropout, McNemar test crosstabulation. (Own creation)

Dropout * Group Crosstabulation

			Group		Total
			Treatment	Control	
Dropout	No drop-out	Count	104	76	180
		% within Group	80,6%	60,8%	70,9%
	Drop-out	Count	25	49	74
		% within Group	19,4%	39,2%	29,1%
Total		Count	129	125	254
		% within Group	100,0%	100,0%	100,0%

Table 6.8 Effectiveness, dropout, Phi and V-Cramer tests. (Own creation)

Symmetric Measures

		Value	Approximate Significance
Nominal by Nominal	Phi	,218	,001
	Cramer's V	,218	,001
N of Valid Cases		254	

The McNemar test as well as the Phi and V-Cramer test confirm the hypothesis that the implementation of an e-learning artefact reduces the drop-out quantity in the usage of SAP S/4 HANA Fiori apps. However, this result addresses the question, if there could be other reasons behind the number of dropout cases. In this sense one of the issues to be analysed is if the employers´ subjective SAP knowledge or the time that the employers took to do the exercises, could play a role for the dropout quantity. Before the employees participated in the experiment, they were asked to fill out a questionnaire. Among several questions, they were asked to estimate how their current SAP ERP knowledge is. For this purpose, they have to respond within a 6 Likert-scale. The value 0 was defined as a very bad SAP knowledge, while 6 was defined as a very good SAP knowledge.

To see the potential relation between the variables, subjective knowledge of SAP and dropouts, the Spearman-Rho-Test was carried out. In this Spearman-Rho-Test an ordinal variable (from 1 to 6) in terms of subjective SAP Knowledge, as well as the dropout variable were compared. The Spearman-Rho-Test shows a negative correlation coefficient of -0.009, being this coefficient very weak (Brosius, 2013, p. 528). Also, a significance of 0.883 is very weak (see Table 6.9).

This means that previous subjective SAP knowledge of SAP ERP does not imply that someone drops out of in the middle of an exercise with SAP S/4 HANA.

Table 6.9 SAP Knowledge and dropout, Spearman-Rho test. (Own creation)

Correlations

			SAPKnowledge	Dropout
Spearman's rho	SAPKnowledge	Correlation Coefficient	1,000	−,009
		Sig. (2-tailed)	.	,883
		N	254	254
	Dropout	Correlation Coefficient	−,009	1,000
		Sig. (2-tailed)	,883	.
		N	254	254

If the subjective SAP Knowledge does not play a role for the dropouts, the next question is if the working time of the employees doing the exercises might have an impact in the dropout cases. For this reason, the control and treatment group in comparison to the variables seconds as well as dropout was analysed with a Mann-Whitney-U test and the Kruskal-Wallis-H test.

Tables 6.10 and 6.12 show that the less time needed to carry out the exercises caused also less dropouts quite significantly. This also can be observed in the lower mean ranks in seconds of the treatment group in comparison to the control group (see Tables 6.11 and 6.13). The reason behind could be that the treatment group could at any time watch the embedded videos in the e-learning artefact, need less time to do the exercises, and thus be more motivated. This is also shown in similar e-learning studies where the Flow Theory played an important role when it comes to learning success (Choi, D.H, Kim, J., Ki, S.H., 2007, 223). According to Csikszentmihalyi´s Flow Theory (1990, p. 6) flow is a state of mind when people's *"consciousness is harmoniously ordered, and they want to pursue whatever they are doing for their own sake"*. In this sense Csikszentmihalyi argues that the experience of flow in the work area has a positive influence in the way people are active and motivated (ibid. p. 158). This could imply also that the treatment group experienced more flow and thus was more motivated to complete the exercise.

Table 6.10 Time and dropouts, Mann-Whitney-U test. (Own creation)

Test Statistics[a]

	Salesorder-seconds	Outbound-seconds	Pickedsec-onds	Billingsec-onds	Newsalesor-derseconds
Mann-Whitney U	5097,000	3457,500	2482,000	1923,500	2829,500
Wilcoxon W	20850,000	19033,500	17882,000	17323,500	17880,500
Z	−2,324	−2,854	−2,003	−2,338	−1,795
Asymp. Sig. (2-tailed)	,020	,004	,045	,019	,073

a. Grouping Variable: Dropout

Table 6.11 Time and dropouts, Man-Whitney-U test. Ranks. (Own creation)

Ranks

	Dropout	N	Mean Rank	Sum of Ranks
Salesorderseconds	No drop-out	177	117,80	20850,00
	Drop-out	71	141,21	10026,00
	Total	248		
Outboundseconds	No drop-out	176	108,14	19033,50
	Drop-out	53	137,76	7301,50
	Total	229		
Pickedseconds	No drop-out	175	102,18	17882,00
	Drop-out	36	124,56	4484,00
	Total	211		
Billingseconds	No drop-out	175	98,99	17323,50
	Drop-out	30	126,38	3791,50
	Total	205		
Newsalesorderseconds	No drop-out	173	103,36	17880,50
	Drop-out	40	122,76	4910,50
	Total	213		

Table 6.12 Time and dropouts, Kruskal-Wallis-H test. (Own creation)

Test Statistics[a,b]

	Salesorder-seconds	Outbound-seconds	Pickedsec-onds	Billingsec-onds	Newsalesorder-seconds
Kruskal-Wallis H	5,401	8,144	4,013	5,464	3,222
df	1	1	1	1	1
Asymp. Sig.	,020	,004	,045	,019	,073

a. Kruskal Wallis Test
b. Grouping Variable: Dropout

Table 6.13 Time and dropouts, Kruskal-Wallis-H test. Ranks. (Own creation)

Ranks

	Dropout	N	Mean Rank
Salesorderseconds	No drop-out	177	117,80
	Drop-out	71	141,21
	Total	248	
Outboundseconds	No drop-out	176	108,14
	Drop-out	53	137,76
	Total	229	
Pickedseconds	No drop-out	175	102,18
	Drop-out	36	124,56
	Total	211	
Billingseconds	No drop-out	175	98,99
	Drop-out	30	126,38
	Total	205	
Newsalesorderseconds	No drop-out	173	103,36
	Drop-out	40	122,76
	Total	213	

6.2.1 Create a Billing and Post the Billing

One of the aspects for effectiveness is also the completion of the exercise step in the process create billing, that also includes a sub-task for posting the billing.

This task was compared for the treatment and the control group. In this case there are on one side a nominal variable (treatment and control group) as well as an ordinal variable (saves-posted-Billing). For this case the Mann-Whitey-U test was utilised. Results show a clear significance (0.001) of the treatment group in comparison to the control group (see Tables 6.14 and 6.16).

Table 6.14 Create and post billing, Man-Whitney-U test, Saved-Post-Billing. (Own creation)

Test Statistics[a]

	SavedPosted-billing
Mann-Whitney U	6593,500
Wilcoxon W	14468,500
Z	−3,272
Asymp. Sig. (2-tailed)	,001

a. Grouping Variable: Group

Table 6.15 Create and post billing, Man-Whitney-U test. Saved-Post-Billing. Ranks. (Own creation)

Ranks

	Group	N	Mean Rank	Sum of Ranks
SavedPostedbilling	Treatment	129	138,89	17916,50
	Control	125	115,75	14468,50
	Total	254		

Table 6.16 Create and post billing, Kruskal-Wallis-H test, Saved-Post-Billing. (Own creation)

Test Statistics[a,b]

	SavedPostedbilling
Kruskal-Wallis H	10,709
df	1
Asymp. Sig.	,001

a. Kruskal Wallis Test
b. Grouping Variable: Group

Also, the mean ranks show better results for the treatment group (see Tables 6.15 and 6.17). This means that the treatment group obtain better results in terms that they could complete the main and sub-task better. This also confirms the hypotheses that the implementation of a e-learning artefact reduces the dropout quantity in the usage of SAP S/4 HANA Fiori apps.

Table 6.17 Create and post billing, Kruskal-Wallis-H test. Saved-Post-Billing. Ranks. (Own creation)

Ranks

	Group	N	Mean Rank
SavedPostedbilling	Treatment	129	138,89
	Control	125	115,75
	Total	254	

6.2.2 Create New Sales Order and Change the Price

In the case of the exercise related to creating a new sales order as well as change the price, the treatment group also performed better than the control group. For this purpose, there is a nominal variable (treatment and control group) and an ordinal variable (create a new sales order and change the price with values 0, 1, 2). In the case of the ordinal variables 0 means that the price is not given, 1 means that the price is the standard price and 2 means that the price was changed to the correct one. Tables 6.18 and 6.20 show the differences between the treatment and the control group were significant (p value = 0.008) in terms that the treatment group performed better (Tables 6.19 and 6.21).

Table 6.18 Create new sales order and change the price, Man-Whitney-U test. (Own creation)

Test Statistics[a]

	CreateNewChangePrice
Mann-Whitney U	6825,500
Wilcoxon W	14700,500
Z	−2,658
Asymp. Sig. (2-tailed)	,008

a. Grouping Variable: Group

Table 6.19 Create new sales order and change the price, Man-Whitney-U test. Ranks. (Own creation)

Ranks

	Group	N	Mean Rank	Sum of Ranks
CreateNewChangePrice	Treatment	129	137,09	17684,50
	Control	125	117,60	14700,50
	Total	254		

Table 6.20 Create new sales order and change the price, Kruskal-Wallis-H test. (Own creation)

Test Statistics[a,b]

	CreateNewChangePrice
Kruskal-Wallis H	7,066
df	1
Asymp. Sig.	,008

a. Kruskal Wallis Test
b. Grouping Variable: Group

Table 6.21 Create new sales order and change the price, Kruskal-Wallis-H test. Ranks. (Own creation)

Ranks

	Group	N	Mean Rank
CreateNewChangePrice	Treatment	129	137,09
	Control	125	117,60
	Total	254	

6.2.3 Non Dropouts and Time

In terms of employees completing the exercises as well as the time needed to complete those, it was analysed if any difference between the treatment group and the control group existed. The dependant variable in this case refers to total seconds needed to complete the exercises. Of 254 participants, 172 could finish the exercise and have the time for all exercises completed. To see if there were any difference between the treatment and the control group for those who have completed the exercises, a t-test was originally thought. Also, the conditions for a t-test as explained in section 6.1 above were checked. The dependent variables values are continuous. Also, the observations are individual and independent, meaning that there is no linkage or direct influence of the values in the observations within the groups. However, whether the existence of outliers or if the dependent variables are normally distributed is something to be checked.

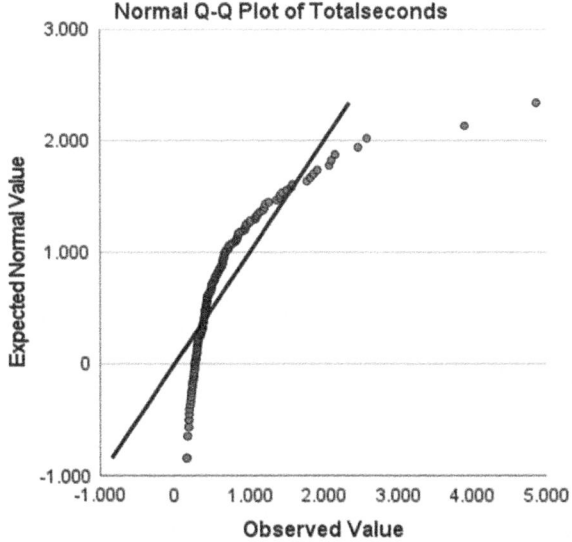

Figure 6.2 Normal Q-Q-Points total seconds. (Own creation)

The Q-Q Plot of point in Figures 6.2 and 6.3 above show that the variables related to the time in total seconds are not normally distributed and have several significant outliers. Also, normal distribution tests based on Kolmogorov-Smirnoff test and the Shapiro-Wilk test were carried out (see Table 6.22). The significance values in the Table 6.22 show below smaller than .001, what means that the variables are not normal distributed (Brosius, F., 2013, p. 405).

Due to the fact the t-test could not be carried out, because the condition of normal distribution cannot be fulfilled, a Mann-Whitney-U test was performed. In the Tables 6.23 and 6.25 it can be seen that the treatment group significantly needed less time that the control group (p value = 0.012) to carry out the exercises. Specifically in Tables 6.24 and 6.26 it can be observed an approximate difference of 25% (24.67%). This would mean that employees needed around 25% less time to complete the exercises. For an international company this could considerably reduce the costs of the training through an e-learning artefact as also stated in other studies (Paa, 2013, p. 98) at universities.

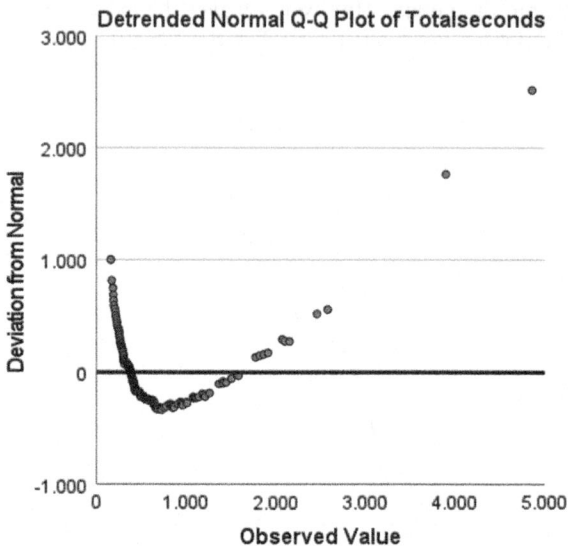

Figure 6.3 Detrended Normal Q-Q-Points total seconds. (Own creation)

Table 6.22 Test of normality. Kolmogorov-Smirnov and Shapiro-Wilk tests. (Own creation)

Tests of Normality

	Kolmogorov-Smirnov[a]			Shapiro-Wilk		
	Statistic	Df	Sig.	Statistic	df	Sig.
Totalseconds	,208	172	,000	,673	172	,000

a. Lilliefors Significance Correction

Table 6.23 Non-dropouts, time, Mann-Whitney-U test. (Own creation)

Test Statistics[a]

	Totalseconds
Mann-Whitney U	2802,000
Wilcoxon W	7752,000
Z	−2,514
Asymp. Sig. (2-tailed)	,012

a. Grouping Variable: Group

Table 6.24   Non-dropouts, time, Mann-Whitney-U test. Ranks. (Own creation)

Ranks

	Group	N	Mean Rank	Sum of Ranks
Totalseconds	Treatment	99	78,30	7752,00
	Control	73	97,62	7126,00
	Total	172		

Table 6.25 Non-dropouts, time, Kruskal-Wallis-H test. (Own creation)

Test Statistics[a,b]

	Totalseconds
Kruskal-Wallis H	6,321
df	1
Asymp. Sig.	,012

a. Kruskal Wallis Test
b. Grouping Variable: Group

Table 6.26 Non-dropouts, time, Kruskal-Wallis-H test. Ranks. Own creation.

Ranks

	Group	N	Mean Rank
Totalseconds	Treatment	99	78,30
	Control	73	97,62
	Total	172	

6.2.4   Regression Analysis and Differences

The results above show that the treatment group could perform better in terms of finalising the exercises as well as complete them in less time. However, the questions addresses if certain independent variables like subjective SAP Knowledge, departments, selling units per country, gender, or location (headquarters and abroad) could make a difference in the results. For example, if the total seconds could be explained by the independent variables above. In this sense the multiple linear regression applies with the formula

$$y = b1 * x1 + b2 * x2 + \ldots + bn * xn + a$$

where n is the number of the independent variables, represented through the variables $\times 1$ to Xn, a is the constant variable and y is the dependent variable (Brosius, F., 2016, p. 448). For these variables, dummy variables were created. While the general subjective SAP Knowledge was based on a formular, the variables related to departments, selling units per country, gender and location were taken objectively from SAP HR module as well as from Microsoft Teams of the company Leica Geosystems AG. In total 14 different variables (some of them dummy variables were selected). These independent variables are treatment and control group, subjective SAP knowledge, gender (female, male), location (headquarters, organisation abroad), department (development and production, logistics, share services and technical service), selling units in the countries USA & Canada, Germany, France, Italy, Spain, UK and Poland, as well as SAP areas of experience. To approach the question related to the differences a regression analysis of these variables with the seconds needed to perform the tasks for the creation of the sales order, the outbound delivery, the pick outbound delivery, the billing request, the new sales order as well as the total seconds were carried out. Based on the regression analysis related to the variables mentioned above, in all the results the R^2 (coefficient of determination) value was lower than 0.2, meaning that the independent variables correlate very weak with each dependent variable (Brosius, F., 2014, p. 283). However, the treatment group as a dummy variable shows in comparison to the control group significance through individual coefficients (see Tables 6.27 and 6.29). None of the variables show any significance in relation to the significant values of the treatment group. Additionally, the Durbin-Watson-Test (see Appendix D in additional electronic material) in all carried multiple regression analysis show a value (between 1.9 and 2.2) close to 2, what means that no correlation exists (Brosius, F., 2016, p. 452). In this sense the questions appear how big the difference between the treatment group and the control group and which factors might influence this difference.

Table 6.27 Multiple linear regression, treatment, and control group. (Own creation)

	Salesorder seconds Sig.	Outbound-delivery seconds Sig.	Picked-seconds Sig	Billing-seconds Sig	Newsales-orderseconds Sig
(Constant)	0,149	0,006	0,071	0,039	0,005
Group = Treatment	0,413	**0,012***	**0,055**	0,187	**0,02***
SAPKnowledge	0,642	0,745	0,814	0,888	0,118
Gender = Female	0,109	0,073	0,655	0,948	0,46
Department = Devprod	0,513	0,165	0,704	0,997	0,743
Department = Logistics	0,166	0,905	0,24	0,107	0,647
Department = Technical- services	0,695	0,562	0,673	0,972	0,488
Land = D	0,186	0,188	0,955	0,384	0,111
Land = USACA	0,933	0,736	0,861	0,633	0,815
Land = ES	0,65	0,397	0,846	0,804	0,636
Land = F	0,825	0,981	0,776	0,592	0,932
Land = PL	0,319	0,152	0,685	0,101	0,507
Land = UK	0,107	0,186	0,068	0,962	0,693
Location = Headquarters	0,231	0,372	0,963	0,366	0,128

Table 6.28 shows this difference in terms of mean seconds between the treatment group and the control group. The treatment group needs significantly less seconds to finish the exercise. While the control group has to read through the notes again and remember the training when doing the exercise, the treatment group can at any time see in form of videos the exercises to be performed. This might explain the better performance of the treatment group. On the other side no other factors in this experiment like subjective SAP Knowledge, gender, department, country, areas of experience and location (headquarters or overseas) are statistically significant. The fact that none of these factors are significant, it explains also that SAP S/4 HANA is completely new for the participants (a new graphical user interface) as well as the exact representation of the process sequences. This explanation speaks positively for the e-learning artefact itself and its pragmatic methodology as well as the didactic theories and the learnings of Design Science Research when developing e-learning artefacts.

Table 6.28 Mean values, treatment, and control group. (Own creation)

Ranks

	Group	N	Mean Rank	Sum of Ranks
Salesorderseconds	Treatment	127	119,79	15213,50
	Control	121	129,44	15662,50
	Total	248		
Outboundseconds	Treatment	119	99,07	11789,00
	Control	110	132,24	14546,00
	Total	229		
Pickedseconds	Treatment	118	91,09	10749,00
	Control	93	124,91	11617,00
	Total	211		
Billingseconds	Treatment	116	89,62	10395,50
	Control	89	120,44	10719,50
	Total	205		
Newsalesorderseconds	Treatment	113	99,54	11247,50
	Control	100	115,44	11543,50
	Total	213		

Table 6.29 Multiple linear regression, treatment and control group, areas of experience. (Own creation)

	Salesorder seconds Sig.	Outbound-delivery seconds Sig.	Picked-seconds Sig	Billing-seconds Sig	Newsales-order-seconds Sig
(Constant)	0,000	0,000	0,000	0,000	0,000
Group = Treatment	0,485	**0,008****	0,088	0,888	0,075
Cservice	0,228	0,197	0,461	0,537	0,541
Warehouse	0,616	0,241	0,159	0,325	0,397
Technicalservice	0,736	0,784	0,405	0,236	0,903
Finance	0,935	0,091	0,284	0,56	0,941
Controlling	0,269	0,051	0,083	0,586	0,313
Humanresources	0,837	0,383	0,45	0,546	0,688
Production	0,416	0,606	0,368	0,694	0,712
Purchasingprocurement	0,879	0,623	0,685	0,708	0,434
Productmanagement	0,633	0,253	0,139	0,34	0,269

6.2.5   Conclusion of the Evaluation Results

The results of the validation of the hypotheses defined in section 5.4.1 are classified in effectiveness and efficiency related results. While the effectiveness is
measured through dropouts in terms that employees could not complete the exercises, the efficiency is related to the time needed in completing every step of the
process (i.e.., create sales order, create outbound delivery, pick outbound delivery,
create invoice and post it, create a new sales order and change the price). The
results of the Kruskal-Wallis-H and Mann-Whitney-U tests show that the treatment group with the e-learning artefact could significantly (see Tables 6.3 and 6.4
to see how big these differences are) finish the exercises related to the process
steps create outbound delivery (p <0.001), pick outbound delivery (p <0.001), and
create invoice (p <0.001). The results of the Mann-Whitney-U test and Kuskal-
Wallis-H test confirm the hypothesis that the implementation of an e-learning
artefact positively influences efficiency utilising SAP S/4 HANA Fiori apps. For
the effectiveness the McNemar test as well as the Phi and V-Cramer show that
there were significantly less dropouts (p <0.001) in the case of the treatment
group (25 cases) in comparison to the control group (49 cases). More specifically the percentage of dropouts for the treatment group sits at 19,4%, while the
control group counts for 39.2% (see Table 6.7). The McNemar Test as well as
the Phi and V-Cramer test confirm the hypothesis that the implementation of an
e-learning artefact reduces the drop-out quantity in the usage of SAP S/4 HANA
Fiori apps. Another aspect is to see if there is a relationship between the subjective SAP Knowledge estimated through the questionnaires and the percentage of
dropouts. To see the potential relation between the variable subjective knowledge
of SAP and dropouts, the Spearman-Rho-Test was carried out. In this Spearman-
Rho-Test an ordinal variable (from 1 to 6) in terms of subjective SAP Knowledge,
as well as the dropout variable were compared. The Spearman-Rho-Test shows a
negative correlation coefficient of -0.010, and a significance of 0.868. This means
that previous subjective SAP knowledge of SAP ERP does not imply that someone drops out of in the middle of an exercise with SAP S/4 HANA. Additionally,
it was important to analyse if other variables like subjective SAP knowledge,
gender (female, male), location (headquarters, organisation abroad), department
(development and production, logistics, share services and technical services),
selling units in the countries USA & Canada, Germany, France, Italy, Spain,
UK and Poland, as well as SAP areas of experience might have an impact on
the results of the experiment. For this purpose, a multiple linear regression was
applied. Results show that none of these factors affect the better performance in

time of the treatment group in comparison to the control group. This drives to the conclusion that an e-learning artefact as a result of the didactic method, the ISSM scientific requirements and the 3-2-1 expositional model is the trigger for the better performance of the treatment group.

Design Guideline for the Conceptualisation and Implementation of an E-Learning Artefact for SAP S/4 HANA

7

After the validation of the hypotheses and the presentation of the concluding results about the better performance of the treatment group with the e-learning artefact, this chapter aims to provide a guideline for any company, who would like to implement such an e-learning artefact. This e-learning artefact can be implemented for SAP S/4 HANA or for other ERP systems. Specifically, it should provide a procedure for any company using an ERP system who is confronted with questions related to effective and efficient ERP trainings for standard, but also continuous improvement of business processes. This guideline is based on the four phases preparation, execution, monitoring, and adaptation. It should provide a sustainable structured procedure in the sense of a platform, upon a company can over time be more efficient in terms of time and costs and be effective in terms of employees' competences acquisition for ERP training purposes.

7.1 Preparation

The preparation phase covers an important area before starting any implementation. It is based on a project planning from the Project Management literature. The idea behind is to have a structured approach because the implementation of the e-learning artefact will influence several areas in an organisation, and it will require several resources. There are many project management methods in the literature. For the implementation of this guideline, it is recommendable to use the project planning of the IPMA (International Project Management Association). IPMA is present in 70 nations with more than 120000 members and its project management certifications are recognised internationally (PMA, 2022, para. 4). Specifically for this chapter the project planning of the Project Management Austria (PMA) as part of the IPMA has been considered (PMA, 2018, p. 5). It covers

© The Author(s), under exclusive license to Springer Fachmedien Wiesbaden GmbH, part of Springer Nature 2023

F. Garayo Maiztegui, *Design and Evaluation of an E-Learning Artefact for the Implementation of SAP S/4 Hana®*, Gabler Theses, https://doi.org/10.1007/978-3-658-40731-5_7

159

the following aspects: project objectives, object structure plan, work breakdown structure, work packages, activity distribution chart, project deadlines, project resources, project financials, financial resources, and project risks (PMA, 2018, p. 5).

There are different types of project objectives that can be considered. They can be content, dates and budget related objectives (PMA, 2018, p. 30). In the case of the project "e-learning artefact" content objectives could refer to the ERP business processes that need to be trained. Further objectives could be related to the prioritisation of business process trainings, to the knowledge competences that employees should acquire, or to the expected quality (PMA, 2018, 30). The object structure plan is related to the material and immaterial objects that can be part of the project like e.g., stakeholders, hardware, and software (PMA, 2018, 31). For example, a mind map can be created with the stakeholders in the sense of HR Training Development department, IT department, SAP training departments. This mind map could also contain the needed LMS software like Chamilo LMS version, as well as hardware like tablets, personal Computers or smartphones, upon the application should run.

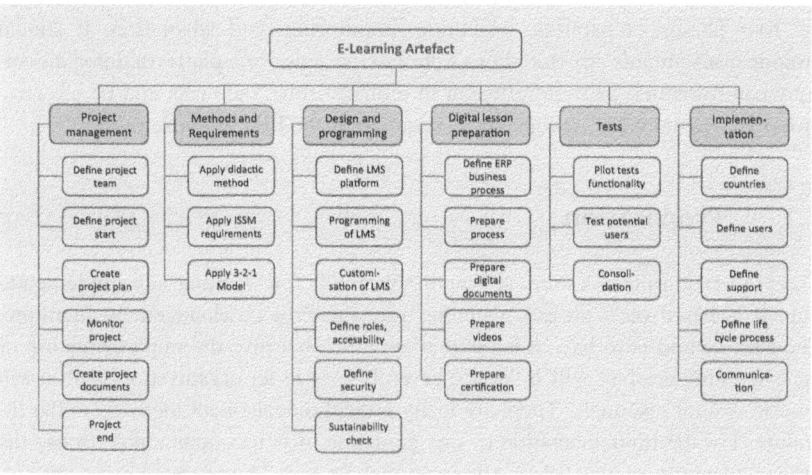

Figure 7.1 Example of a project structure plan for the e-learning artefact project

Also, security technologies like VPN can be considered. The project structure plan should provide the working packages that are needed to carry out the project

(Patzak, & Rattay, 2014, p. 223). These work packages (see Figure 7.1) follow a chronological structure to make sure that the co-ordination between the different work packages take place. After the work packages have been created, the next step is the activity distribution chart. This activity distribution chart should contain a detailed allocation of work packages to different project stakeholders in the sense of roles in the organisation (PMA, 2018, p. 35). This means that work packages should be assigned to members of the IT department, the HR department of the SAP support department. At this stage it is important to ensure that the stakeholders do have the needed competences to carry out the project. For example, the IT programmers should have the competence to install, but also customise the LMS platform. Based on the project structure work packages, a project deadlines chart with the milestones is needed. The project chart should name the work packages in a timeline and show how these packages interact sequentially with each other. A project will normally imply internal and/or external resources. In the case of internal resources, the project should contain personnel's man days for each part of the work packages. The project costs are based on internal and external resources. While the internal resources will multiply the number of man days by the internal daily cost rate, the external costs will include the costs for carrying out certain tasks of the project. Some external companies quote the investment as a flat cost, while others prefer to name the number of days and the charge rate per day. In general, it can be said, that the project planning should be the basis to get an overview about what a company wants to achieve with an e-learning artefact for a sustainable ERP training, identify and ensure that the resources in terms of capacities and capabilities are available, and provide a reliable information for the company's top management. The project planning could be the basis for the approval of the project.

7.2   Execution

The execution part in this chapter will describe only certain work packages that are crucial for the implementation of the e-learning artefact validated in this research project (see Figure 6.3). Once a proper project planning exists; the next key work package is the implementation of the "methods and requirements" as described in section 5.2. Thus, it is important to define, based on perspective schema for the teaching planning, the conditions of the learning, the context reasons (what is the relevance for the present and the future, what is the exemplary meaning), the thematical structure (learning objectives, proof, and monitoring), the presentation possibilities (3-2-1 expositional model) and the methodical structure (structure of course like in Table 5.8). Additionally, it should be considered the requirements of the ISSM model when developing an e-learning artefact (see section 5.2) as well as the requirements of the company. The work package "methods and requirements" is the grounded basis for an effective and efficient ERP training. The didactic methods and ISSM requirements upon an e-learning artefact is built-up, makes an ERP training effective and efficient, not the LMS e-learning tool itself. In fact, there are many e-learning artefacts on the shelf, however these are mostly grounded only in the availability of technical possibilities (e.g., including videos, etc.).

The work package "design and programming" is the development of the e-learning artefact itself. Before the development takes place, a technical specification concept should provide an overview of what is needed. This technical specification concept should not only be related to the e-learning artefact itself, but also it should consider the technical requirements when creating the digital documentation (e.g., videos) of a course. For example, if the videos have a mp4 format, the embedment of this videos in the e-learning artefact should be integrated in a way, that the videos can be played first time. The technical specification concept should also include in which browsers the e-learning artefact will run. Sometimes browsers technical requirements do not always match with the LMS current technical possibilities. This will reduce the risks of surprises in terms of technical requirements and will help with the selection of a proper LMS. In this research project the LMS Chamilo, among several LMS systems available in the market, was selected, because it contain features needed from the ISSM requirements. This work package should also consider the IT security. At this stage it is also important to define which employees´ roles have access to which contents, as well as which security log-in mechanisms need to be set up (e.g., VPN connection or single sign on accessibility). There should be a clear accessibility concept in place to ensure that sustainability of the e-learning artefact as a platform for all ERP trainings of the company is valid now and in the future. The "design and programming" package should also consider corporate identity. When employees log into a system and see that the design corresponds to the corporate identity, they will identify themselves more with the e-learning artefact. If corporate identity guidelines are held on, employees will feel a part of it. This could also increase the acceptance and motivation towards the usage of the e-learning artefact.

The work package "digital lesson preparation" refers to the content of the ERP business process lesson. It should be based on a clear structure (see Figure 5.12 in section 5.2). It should include first the ERP business process workflow with an

explanatory video. Learners should according to APO-IT subject didactic understand, what are the steps of the process to be carried out. This will give them a
sense of knowledge abstraction independent of the ERP system. Later the users
should be able to view an overview of these steps with the software also with an
explanatory video, e.g., which Fiori app or function key corresponds with part
of the process. Learners will then be able to orientate themselves in the software. Next the log-in or link with the ERP system is needed. Learners will get
confident that they can access the system and that it works. This would mean a
first success experience for them. Afterwards, every process step should contain
in form of a video the detailed activities to process a step. The video will give
provide them the possibility to see it as much as they want and thus have the
confidence that the can make the exercise. An exercise to learn proactively this
step is also included, so that learners can experience in the ERP system how the
step is carried out.

Testing an existing system is crucial to ensure that quality is given, and the
learners can carry out the exercises without technical problems. As it was shown
in section 4.4 users expect that the e-learning artefact runs without any application
errors, and it is always available (system quality). This is a very crucial part that
can make the difference when implementing the artefact in terms the acceptance
and utilisation of the e-learning artefact. If learners do not have any problems
using the e-learning artefact, they will experience more flow. In this way they
will be consciously absorbed by the course itself and will not need to worry
about technical issues that limits the utilisation of the e-learning artefact. Apart
of the technical tests, these should also include the feedback of 2 or 3 persons
related to the content. Is the content of the video understood? or are there spelling
mistakes on it?, etc. Feedback related to these kinds of questions are needed to
ensure that the content quality is acceptable.

Once the tests are successful, the company can then start with the implementation for e-learning purposes. There should be an official communication to the
employees by the top management, so that employees can see the importance of
the e-learning artefact. Also, participants should receive an invitation per email
with the e-learning course. In this invitation participants should understand why
they have been invited, what is the content of the course, what do they get by
making the course (e.g., certificate of achievement), until when the course is supposed to be completed, a link with the e-learning course and support contact
details in case they would have technical problems. In case learners have issues
with the e-learning artefact, there should be a strong support team who is able to
reply and solve these issues ad-hoc and in an amiable way. This communication
effort is quite time consuming, but at the same time an important contributor for

the acceptance of the e-learning artefact. An initial successful implementation is not the end of the implementation of an e-learning artefact. Every process is submitted to changes and adaptations. These changes can be triggered by new market or product requirements. Thus, it is important to establish a life cycle process, upon which an existing e-learning course can be changed and published in the e-learning artefact. In this sense it is recommendable to use SAP EnableNow as a tool. SAP EnableNow allows to make changes of this process at specific places and then publish the documentation as well as the videos. It is one of the most efficient ways that I experienced to change and adapt documentation. This makes the trainings of SAP business process sustainable in terms of effectiveness and efficiencies and reduce the costs of trainings considerably. SAP EnableNow can be used for SAP S/4 HANA as well as for other ERP systems.

7.3   Monitoring

The phase monitoring is essential to understand if the learning objectives have been achieved. One possibility is to see in the ERP system directly if the participants have completed the exercises according to the e-learning training. Based on ERP reports, it can be analysed the quality as well as the completion of the exercises. Also, the lead time needed between the completion of the first and the last exercise can be measured. Upon completion of the exercises, it is recommendable to provide learners with a participation certificate. This follows two main objectives. On one side participants get rewarded with a certificate, and on the other side a company can identify which employees have which competences in SAP. The monitoring of the SAP competences of employees should be part of the strategic personnel development in a company. This is especially crucial when new employees join the company and overtake certain SAP roles. Thus, a company can define which trainings with the e-learning platform based on which roles a new employee needs to complete. The same applies for the further development of employees in new SAP roles.

7.4   Adaptation

The adaptation phase corresponds to the life cycle of the e-learning artefact. During the preparation phase, as part of the project structure plan, the implementation includes the life cycle process. The life cycle process establishes the procedure to make changes in the digital courses embedded in the e-learning artefact. Using

SAP EnableNow it is quite easy to make a change at a certain step of the process. This change may not only contain the adaptation of a SAP business process file or the change of a text, but also the change of sound at a specific sequence; i.e., a voice can be additionally be added. After the change is done, SAP EnableNow can save the documentation e.g., in a Word, PowerPoint of pdf file. Additionally, a new video can be created that replaces the old one. This means that very few efforts are needed, without the need of making a whole new documentation or record a whole new video from the very beginning again. SAP EnableNow can be used to record SAP S/4 HANA software, but also other type of ERP software. The savings in this sense are considerably high. These are benefits for companies that currently are reluctant to invest in digital ERP trainings.

Conclusion 8

This research project aimed to create a new e-learning artefact for SAP S/4 HANA training purposes. Based on Design Science Research, a new prototype of an e-learning artefact was developed. The three-cycle view of Design Science Research was the basis for the structured approach of the environment, knowledge base and design cycle. The environment refers to the requirements of the company Leica Geosystems as well as the SAP S/4 HANA business processes. The knowledge base is the basis for the scientific theories and methods, upon which the e-learning artefact is built-up. On one side the result of the knowledge base is a didactic methodical concept as a consolidation from Klafki's perspective schema from the educational theory, the Heimann's Hamburger model from the learning theory, the APO-IT subjective didactic and Kerre's 3-2-1 expositional model. On the other side the knowledge base also considers Delones & McLeans' Information System Success Model (ISSM) relevant for the requirements of a successful e-learning artefact. The technical development of the e-learning artefact was carried out with the learning management system (LMS) Chamilo, an open-source program that enable several perspectives. The videos and other training materials were developed with SAP EnableNow. The usage of Chamilo allows the creation of a sustainable platform upon which a company can grow in terms of conceptual trainings systems. The software SAP EnableNow makes possible that trainings can technically be built-up in an efficient way. This is applicable for standard SAP business processes, as well as customised and adapted SAP business processes. One of the key aspects of Design Science Research is the *research rigor*, this means, the rigor utilizing scientific methods when creating and evaluating artefacts (Hevner et al., 2004, p. 87). The scientific methods were covered by the three-cycle Design Science Research. The evaluation of the artefact was carried out with an experimental research design with a treatment and a control group, and with a sample size of 254 employees. While the treatment

© The Author(s), under exclusive license to Springer Fachmedien Wiesbaden GmbH, part of Springer Nature 2023

F. Garayo Maiztegui, *Design and Evaluation of an E-Learning Artefact for the Implementation of SAP S/4 Hana®*, Gabler Theses, https://doi.org/10.1007/978-3-658-40731-5_8

did the course using the e-learning artefact, the control group received conventional online trainings. The experiment was carried from March 2021 until January 2022. The results of the experiment were classified in effective and efficient results. While the effectiveness was measure with the number of dropouts (employees not completing the exercise), the efficiency was measured in terms of time.

For the effectiveness the McNemar test as well as the Phi and V-Cramer show that there were significantly less dropouts (p <0.001) in the case of the treatment group (25 cases) in comparison to the control group (49 cases). More specifically the percentage of dropouts for the treatment group sits at 19,4%, while the control group counts for 39,2% (see Table 6.7). The results of the Kruskal-Wallis-H and Mann-Whitney-U tests show that the treatment group with the e-learning artefact could significantly (see Tables 6.3 and 6.5 to see how big these differences are) finish the exercises quicker in terms of the process steps create outbound delivery (p <0.001), pick outbound delivery (p <0.001), and create invoice (p <0.001). The results of the Mann-Whitney-U test and Kuskal-Wallis-H test confirm the hypothesis that the implementation of an e-learning artefact positively influences efficiency utilising SAP S/4 HANA Fiori apps. This research project contains at the end a guideline how to implement such a learning artefact for SAP S/4 HANA, and also for other type of ERP systems. It gives an overview about the phases, as well as about the technology that should be considered. The idea is that several companies can profit from this study.

This study addresses limitations of other research projects related to the implementation of an e-learning artefact for ERP systems in the past. For example, the literature review for this study concludes that most of the studies were done at universities with students. Studies carried out in a high school or in companies may vary the outcomes and learning self-efficacy considerably due to different learning environments as well as the age of the individuals (Chu, 2010, p. 255). Additionally, the only studies that were validated at companies had a quite small sample (16 as well as 52 adults). Also, these studies were carried out in one single country, which limits the experiment into one culture. Further, most of the studies were based on surveys through the collection of data through questionnaires, which makes the studies based on subjective perceptual data, and less on objective measures like tests and grades. This study was validated with an experiment, where 254 employees of the company Leica Geosystems participated. The data of the experiment were taken objectively from the SAP S/4 HANA system. The employees, that participated in the experiment, live and work in several European countries, as well as USA and Canada, with different country cultural backgrounds. However, this study has also some limitations. The study was carried out only at the company Leica Geosystems AG. Further studies should be carried out in different organisations and industries with different company cultures that could eventually influence the acceptance of the e-learning artefact. Additionally, the experiment also considered a posttest-only control group. Further studies could in the research design include two or more control groups. For example, a group with conventional online training, and a group with offline

© The Author(s), under exclusive license to Springer Fachmedien Wiesbaden GmbH, part of Springer Nature 2023

F. Garayo Maiztegui, *Design and Evaluation of an E-Learning Artefact for the Implementation of SAP S/4 Hana®*, Gabler Theses, https://doi.org/10.1007/978-3-658-40731-5_9

training. Also, an experiment could have been carried out with other ERP systems like Microsoft Dynamics. This study is also limited by the availability of resources in terms of funding and time disposability of employees. Finally, it can be said that this study contributes to the scientific world in terms of developing e-learning artefacts with Design Science Research as part of the interventionist research approach, and also to the pragmatic world in terms of providing effective and efficient e-learning solutions for companies that are currently implementing or will implement ERP systems in the future.

Literature

Aff, J. (1996). Wissenschaftsorientierung und Praxisbezug (Situationsorientierung) als curriculare und fachdidaktische Herausforderung für kaufmännische Sekundarschulen. *Festschrift Wilfried Scheider,* pp. 343–371.

Aff, J. (1993). Handlungsorientierung—Mythos oder (wirtschafts)didaktische Innovation. In: Scheider, W. (Hrsg.): komplexe Methoden im betriebswirtschaftlichen Unterricht. Festschrift für Hans Krasensky. Wien 1993, 195–271.

Ajzen, I., & Fishbein, M. (1980). *Understanding attitudes and predicting social behaviour. Englewood Cliffs,* NJ: Prentice-Hall Inc.

Alavi, M., and Leidner, D. E. "Knowledge Management and Knowledge Management Systems: Conceptual Foundations and Research Issues," *MIS Quarterly 25*(1), March 2001, 107–136.

Alcivar, I., Abad A.G. (2016). *Design and evaluation of a gamified system for ERP training.* Computers in Human Behaviour, 58, 109–118.

Analytic steps (2023). *Data Science.* Retrieved February 09, 2023, from https://www.analyt icssteps.com/blogs/greenfield-brownfield-and-bluefield-approach-sap-s4hana

Angolia, G.M., Pagliari, R.L. (2018). Experiential learning for logistics and supply chain management using a SAP ERP software simulation. *Decision Sciences, Journal of innovative education, 16*(2), 104–125.

Arnold, R., Krämer-Stürzl, A., Siebert, H. (1999). *Dozentenleitfaden. Planungs- und Unterrichtsvorbereitung in Fortbildung und Erwachsenbildung.* Berlin, Germany: Cornelsen Verlag.

Ates, N. (2017). *Entwicklung einer Design Science Theorie für Ambient and Assisted Living (AAL) Systeme.* Ein Anwendungsbeispiel auf Grundlage einer Assistenzlösung für das Risiko- und Notfallmanagement in einer institutionellen Pflege (unpublished doctoral thesis), University Innsbruck, Austria.

Atteslander, P. (2010). *Methoden der empirischen Sozialforschung.* Berlin: Erich Schmidt Verlag.

Baard, V. (2010). *A Critical review of interventionist research.* Qualitative Research in Accounting & Management, 7 (1), 13–45.

Babaian, T., Xu, J., Lucas, W. (2017). ERP prototype with built-in task and process support. *European Journal of Information Systems, 27*(2), 189–206.

© The Editor(s) (if applicable) and The Author(s), under exclusive license to Springer Fachmedien Wiesbaden GmbH, part of Springer Nature 2023
F. Garayo Maiztegui, *Design and Evaluation of an E-Learning Artefact for the Implementation of SAP S/4 Hana®*, Gabler Theses,
https://doi.org/10.1007/978-3-658-40731-5

Baumann, R. (1996). *Didaktik der Informatik* (2nd ed.). Stuttgart, Germany: Ernst Klett Verlag Gmbh.

Berg, B., Silvia, P. (2013). *Einführung in SAP HANA.* Germany: Beltz Bad Langensalza GmbH.

Brand, T. (2014). *Evaluation einer arbeitsprozessorientierte IT-Weiterbildung: „IT-Spezialisten"* (Unpublished doctoral thesis), University of Erfurt, Germany.

Brenner, W., Hess. T. (2012). *Wirtschaftsinformatik in Wissenschaft und Praxis:* Festschrift für Hubert Österle. Berlin, Germany: Springer Gabler.

Brosius, F. (2013). *SPSS 21.* Heidelberg: mitp Hüthig, Jehle, Rehm GmbH.

Brosius, F. (2014). *SPSS 22 für Dummies.* Germany: 2014 WILEY-VCH Verlag GmbH & Co. KGaA.

Brosius, H-B., Haas, A., Koschel, F. (2012). *Methoden der empirischen Kommunikationsforschung. (6 ed.).* Munich, Germany: Springer Verlag.

Brenckmann, I., Pöhling, M. (2013). *The SAP HANA Project Guide.* Germany: Expresso Tutorials GmbH.

Buechler, D. (2018). Purchase to pay process. Retrieved June 17, 2022, https://geowiki.lgs-net.com/HGS/Processes/PurchaseToPayProcess.

Bühl, A. (2016). *SPSS 23. Einführung in die moderne Datenanalyse.* 15th ed. Germany: Pearlson Deutschland GmbH.

Burgdorf, J., Destradi, M., Kiss M, Schuber, M. (2017): *Logistik mit SAP S/4 HANA.* Germany: Rheinwerk Publishing.

Burrell, G., Morgan G. (1979). *Sociological Paradigms and Organizational Analysis.* London: Heineman.

Carlsson, A.S. (2006). Design Science Research in Information Systems: A Critical Realist Perspective (2006). *ACIS 2006 Proceedings,* 40. https://aisel.aisnet.org/acis2006/40.

Carlsson, S. A. (2007). Developing Knowledge through IS Design Science Research. *Scandinavian Journal of Information Systems, 19* (2), 75–86.

Charland, P., Cronan, T.P., Léger, P-M., Robert, J. (2015). Developing and assessing ERP competencies: basic and complex knowledge. *Journal of Computer Information Systems, 56*(1), 31–39.

Chauhan, S., Jaiswal, M. (2016). Determinants of acceptance of ERP software training in business schools: Empirical investigation using UTAUT model. *The International Journal of Management Education,* 14, 248–262.

Choi, D.H., Kim, J., Kim, S.H. (2007): ERP training with a web-based electronic system: the Flow Theory perspective. *International Journal of Human-Computer Studies, 65*(3), 223–43.

Chu, R.J. (2010). How family support and Internet self-efficacy influence the effects of e-learning among higher aged adults—Analyses of gender and age differences. *Computers and Education, 55*(1), 255–264.

Conroy, G. (2012). *Business simulations applied in support of ERP training* (Doctoral Thesis). Available in ProQuest. (UMI number 3534618)

Cresswell, J.W. (2014). *Research design: qualitative, quantitative, and mix methods approaches.* (4th ed.). Los Angeles: Sage.

Cronan, T.P., Douglas, D.E. (2011). A student's ERP simulation game: a longitudinal study. *Journal of Computer Information Systems, 53*(1), 4–13.

Cronan, P.T., Léger, P.M., Robert, J., Babin, G., Charland, P. (2012). *Comparing Objective Measures and Perceptions of Cognitive Learning in an ERP Simulation Game: A Research Note.* Simulation & Gaming, *43*(4), 461–480.

Cronan, P.T.., Douglas,, D.E.., Alnuaimi, O., Schmidt, P.O. (2011). Decision Making in an Integrated Business Process Context: Learning Using an ERP Simulation Game. *Decision Sciences Journal of Innovative Education, 9*(2), 227–234.

Csikszentmihalyi, M. (1990). *Flow: the psychology of optimal experience.* New York: Harper & Row.

Czarnocha, B. & Prabhu, V. (2004). *Teaching-research and the design experiment—Two methodologies for the Integration of Research and Classroom Practice.* Retrieved from Homi Bhabha Centre for Science Education website: http://www.hbcse.tifr.res.in/episteme/episteme-1/allabs/prabhuabs.pdf

Cube von, F. (1999). Die kritisch-kommunikative Didaktik. In Gudjons, H., Winkel, R. (Ed.), *Didaktische Theorien (p. 57–74).* Hamburg, Germany: Bergmann und Helbig.

Darban, M., Kwak, D-H., Deng, S., Srite, M., Lee, S. (2016). Antecedents and consequences of perceived knowledge update in the context of an ERP simulation game: A multi-level perspective. *Computers & Education, 103*, 87–98.

Das Statistik-Portal. Statistiken zur SAP SE. (2017). Retrieved November 28, 2017, from https://de.statista.com/themen/232/sap/.

Davis, F.D. (1989). Perceived Usefulness, Perceived Ease of Use, and User Acceptance of Information Technology. *MIS Quarterly, 13*(3), 319–340.

Davis, F. D., Bagozzi, R. P., & Warshaw, P. R. (1989). User acceptance of computer technology: A comparison of two theoretical models. *Management Science, 35*(8), 982–1003.

DeLone, H.W., McLean R.E. (1992). Information Systems Success: The Quest for the Dependent Variable. *Information Systems Research, 3*(1), 60–95.

DeLone, W. H., & McLean, E. R. (2003). The DeLone and McLean model of information systems success: A ten-year update. *Journal of Management Information Systems, 19*(4), 9–30.

Deranek, K., McLeod, A., Schmidt, E. (2019). ERP Simulation Effects on Knowledge and Attitudes of Experienced Users. *Journal of Computer Information Systems, 59*(4), 373–383.

Destradi, M., Kiesel, S., Lorey, C., Schütte, S. (2019). *Logistik mit SAP S/4 HANA.* Germany: Rheinwerk Verlag.

Diekmann, A. (1998). *Empirische Sozialforschung. Grundlagen, Methoden, Anwendungen.* Hamburg: rowohlts enzyklopädie.

Dorobăt, I. (2014). Models for Measuring Success in Universities: A Literature Review. *Informatica Economică vol. 18*(3), 77–90.

Dresch, A., Pacheco D., Valle Antunes, J.A. (2015). *Design Science Reseach. A method for Science and Technology Advancement.* London, UK: Springer Verlag.

Dumay, J., & Baard, V. (2017). An introduction to interventionist research in accounting. In *Hoque,Z., Parker,L.D., Covaleski, M.A. & Haynes K., The Routledge Companion to Qualitative Research Methods*, 265–283. New York: Routledge.

Eberle, F. (1996). Didaktik der Informatik bzw. einer informations- und kommunikationstechnologischen Bildung auf der Sekundarstuffen II. Aarau, Switzerland: Sauerländer Verlag.

Edmondson, A. C., & McManus, S. E. (2007). Methodological fit in management field research. *The Academy of Management Review, 32*(4), 1155–1179.

Elsevier (2021). Measuring a journal's impact. Retrieved from https://www.elsevier.com/aut hors/tools-and-resources/measuring-a-journals-impact

ERPSim (n.d). ERP Sim Lab, Retrieved January 21, 2022, from https://erpsim.hec.ca/en/ erpsim.

Euler, D. & Hahn, A. (2004). *Wirtschaftsdidaktik*. Germany: Haupt UTB.

Fischer, C. (2011). The information systems design science research body of knowledge—a citation analysis in recent top-journal publications. *Proc. of PACIS, Brisbane, Australia,* 1–12.

Flechsig, K.-H., Haller, H.-D (1975). *Einführung in didaktischen Handel. Ein Lehrbuch für Einzel- und Gruppenarbeit*. Germany: Klett Ernst Verlag GmbH.

Frötschl, C. (2015). Enterprise Resource Planning Systeme im kaufmännischen Unterricht. *Schriften aus der Fakultät Sozial- und Wirtschaftswissenschaften der Otto-Friedrich-Universität Bamberg*. Band 20.

Gadatsch, A. (2015). *Geschäftsprozesse analysieren und optimieren. Praxistools zur Analyse, Optimierung und Controlling von Arbeitsabläufen*. Germany: Springer Vieweg Verlag.

Gagné, R.M. (1985). *The conditions of learning and theory of instruction*. New York: CBS College Publishing.

Galvan J.L. (2017). *Writing Literature Reviews. A Guide for Students of the Social and Behavioural sciences*. (6 ed.) London: Routledge.

Goldkuhl, G. (2011). Design Research in Search for a Paradigm: Pragmatism Is the Answer. In *M. Helfert and B. Donnellan. EDSS 2011, CCIS 286*, 86–95. Heidelberg: Springer Verlag.

Goff, M.W., Getenet, S. (2017). Design-Based Research in Doctoral studies: adding a new dimension to Doctoral Research. *International Journal of Doctoral Studies, 12*, 109–119.

Göschl, G. (2002). *Knowledge Management mit SAP R/3 ® knowledge warehouse 5.0* (Unpublished master thesis), University of Economics in Vienna.

Google search engine (2017). Internet. Retrieved October 11, 2017, from https://www.goo gle.com.

Google Books search engine (2017). Internet. Retrieved October 11, 2017, from https:// books.google.com.

Gough, D., Oliver S., Thomas, J. (2012). *An Introduction to Systematic Reviews*. London: Sage.

Gravill, J., Compeau, D. (2008). Self-regulated learning strategies and software training. *Information & Management, 45*, 288–296.

Gregor, S. (2007). The Anatomy of Design Theorie, *Journal of the Association for Information Systems, 8*(5), 312–335.

Gregor, S., Hevner, A. (2013). Positioning and representing Design science research for the maximum impact. *MIS Quarterly, 37*(2), 337–355.

Gudjons, H., Winkel, R. (1999). *Didaktische Theorien*. Hamburg: Bergmann und Helbig Verlag.

Gudjons, H. (2008). *Handlungsorientiert lehren und lernen. Schüleraktivierung, Selbsttätigkeit, Projektarbeit*. Germany: Klinkhardt Verlag.

Gudjons, H. (2008b). *Pädagogisches Grundwissen. Überblick—Kompendium—Studienbuch*. Germany: Klinkhardt Verlag.

Hadaya, P. & Pellerin, R. (2008). Determinants of e-collaboration: Evaluating the intent of Canadian manufacturing films to share inventory information with their suppliers. International journal of eCollaboration, *4*(2), 29–54.

Hasenfeld, Y. and Furman, W. (1994), Intervention research as an interorganizational exchange. In *Rothman, J. and Thomas, E., Intervention research: Design and development for human service*, 297–311. New York: Haworth Press.

Hassanzadeh A., Kanaani, F., Elahi, S. (2012). A model for measuring e-learning systems success in universities. *Expert Systems with Applications*, 39, 10959–10966.

Heričko, M., Rajšp, A., Horng-Jyh, P.W., Beranič, T., (2017). Using a Simulation Game Approach to Introduce ERP Concepts—A Case Study. In *KMO 2017, CCIS 731*, 119–132. Switzerland: Springer International Publishing AG.

Herrington, J., Montgomerie, C. (Ed.), McKenney, S., Seale, J. (Ed.), Reeves, T. C., & Oliver, R. (2007). Design-based research and doctoral students: Guidelines for preparing a dissertation proposal. *World Conference on Educational Multimedia, Hypermedia & Telecommunications*, 4089–4097.

Hwang, M., Cruthirds, K. (2017). *Impact of an ERP simulation game on online learning*. International Journal of Management Education, 15, 60–66.

Hevner, A., March, S., Park, J., & Ram, S. (2004). Design Science in Information Systems Research. *MIS Quarterly, 28*(1), 75–105.

Hevner, A., & Gregor, S. (2013). Positioning and presenting design science research for maximum impact. *MIS Quarterly, 37*(2), 337–355.

Hevner, A. (2007). The three Cycle View of Design Science Research. *Scandinavian Journal of Information Systems, 19*(2), 87–92.

Hevner, A., Chatterjee, S. (2010). *Design research in Information Systems*. New York: Springer Verlag.

Hubwieser, O. (2007). *Lernpsychologische Fundierung*. Germany: Springer Verlag.

Jank, W. & Meyer, H. (1991). *Didaktische Modelle*. Frankfurt a. M.: Cornelsen Scriptor.

Jank, W. & Meyer, H. (2002). *Didaktische Modelle*. Germany: Cornelsen

Jönsson, S., Lukka, K. (2007). There and back again. Doing IVR in management accounting. In *Chapman, C., Hopwood, A., Shields, M., Handbook of Management accounting research, 1, 373 -397*. Elsevier: Amsterdam.

Jesson J.K., Matheson, L., Macey, M.F. (2011). Doing Your Literature Review. Traditional and Systematic Techniques. London: Sage.

Kansanen, P. (1995c). The *Deutsche Didaktik. Journal of Curriculum Studies, 27*(4), 347–352.

Karaali, D., Altin Gumussoy, C., Calisir, F.(2011). Factors affecting the intention to use a web-based learning system among, blue-collar workers in the automotive industry. *Computers in Human Behavior, 27*, 343–354.

Karlsruhe Institut für Technologie. Karlsruher Virtueller Katalog (2017). Retrieved October 11, 2017, from http://kvk.bibliothek.kit.edu.

Kauffeld, S. (2010). *Nachhaltige Weiterbildung*. Germany: Springer Verlag.

Kennedy-Clark, S. (2013). Research by Design. Design-based Research and the Higher Degree Research student. *Journal of Learning Design, 6*(2), 28–29.

Kerres, M. (2018). Mediendidaktik. Konzeption und Entwicklung digitaler Lernangebote. Berlin, Germany: De Gruyter.

Klafki, W. (1999). Die bildungstheoretische Didaktik im Rahmen kritisch-konstruktiver Erziehungswissenschaft. In Gudjons, H., Winkel, R. (Ed.), *Didaktische Theorien* (p. 93–112). Hamburg, Germany: Bergmann und Helbig.

Klafki, W. (1985). *Neue Studien zur Bildungstheorie und Didaktik. Beitrag zur kritisch-konstruktiven Didaktik*. Weinheim, Germany: Beltz.

Kohnen, M. (2011): *Individualisierendes Lehren und Lernen anhand einer multimedialen Lernumgebung zum Thema Sonnenschutz.* Eine Design-Based Research Studie. Germany: Duisburg-Essen Verlag.

Koglin, U. (2016). *SAP S/4 HANA. Voraussetzungen, Nutzen, Erfolgsfaktoren.* Germany: Rheinwerk Verlag GmbH.

Köppl, G., Thallinger, M. (2003). *Grundlagen und Implementierung des ERP basierenden Softwaresystems SAP R/3 am Beispiel eines virtuellen Unternehmens* (Unpublished master thesis), University of Vienna.

Körsgen, F. (2001): *Handlungsorientierte computerstützte Lehr-, Lernarrangements am Beispiel SAP R/3.* Germany: Josef Eul Verlag.

Korte, R. & Mercurio, Z.A. (2017). Pragmatism and human resource development: Practical foundations for research, theory and practice. *Human Resource Development Review, 16* (1), 60–84.

Kron, F.W. (2000): *Grundwissen Didaktik.* Germany: Reinhardt Verlag.

Lave, J., Wenger, F. (1991). *Situated learning: Legitimate peripheral participation.* New York: Cambridge University Press

Lee, J.K., Lee, W.K. (2008). The relationship of e-Learner's self-regulatory efficacy and perception of e-learning environmental quality. *Computers in Human Behaviour*, 24, 32–47.

Lehner, M. (2009): *Allgemeine Didaktik.* Germany: Haupt Berne.

Li, D.C., Tsai, C.Y. (2020). An Empirical Study on the Learning Outcomes of E-Learning Measures in Taiwanese Small and Medium-Sized Enterprises (SMEs) Based on the Perspective of Goal Orientation Theory. *Sustainability (MDPI)*, 12, 5054.

Littell H.A., Corcoran, J., Pillai, V. (2008). *Systematic Reviews and Meta-Analysis.* New York: Oxford University Press.

Livari, J. (2007). A Paradigmatic Analysis of Information Systems as a Design Science. *Scandinavian Journal of Information Systems: 19* (2), Article 5.

Massey University (2021). *Using Scopus and SJR to find a Journal's Impact and Rank.* Retrieved from https://www.massey.ac.nz/massey/research/library/library-ser vices/research-services/publish/ranking-impact-scopus.cfm

Mandl, H., Spada, H. (1988). *Wissenspsychologie.* Germany: Psychologie Verlags Union.

Manson, N.J. (2006): is operations research really research? *Orion*, 22(2), 155–180.

Martial, I. von (2002). *Einführung in didaktische Methode.* Germany: Schneider Verlag Hohengehren.

Mathes, C. Wirtschaft unterrichten. Methodik und Didaktik der Wirtschaftslehre. Germany: Haan-Gruiten.

Matošková, J. (2016). Measuring Knowledge. *Journal of Competitiveness, 8*(4), 5–29.

McKay, J., Marshall, P. (2005). A Review of Design Science in Information Systems. *ACIS 2005 Proceedings.*

McKay, J., P. Marshall, and G. Heath (2010)., An exploration of the concept of design in information systems in Information Systems Foundations. In hart, D.N., Gregor, S. (2010). *The Role of Design Science.* Australia: ANU Press.

McNeill, P., & Chapman, S. (2005). *Research methods.* (2nd ed.). London: Routledge.

Meier, R. (2006). *Praxis E-learning. Grundalgen, Didaktik, Rahmenanalyse, Medienauswahl, Qualifizierungskonzept, Betreuungskonzept, Einführungsstrategie, Erfolgssicherung.* Offenbach, Germany: Gabal Verlag GmbH.

Meyer, H. (1980). *Leitfaden zur Unterrichtsvorbereitung.* Germany: Cornelsen Scriptor Verlag.

Meyer, H. (1987). *Unterrichtsmethoden II.* Germany: Cornelsen Scriptor Verlag.

Möller, C. (1999). Die curriculare Didaktik. In Gudjons, H., Winkel, R. (Ed.), *Didaktische Theorien* (pp. 75–93). Hamburg, Germany: Bergmann und Helbig.

Monk, F.E., Guidry, R.K., Pusecker L.K., Ilvento, T.W. (2019). Blended learning in computing education: It's here but does it work? *Education and Information Technologies,* 25, 83–104.

Mullins, J.K., Cronan, T.P. (2021). Enterprise systems knowledge, beliefs, and attitude: A model of informed technology acceptance. *International Journal of Information Management,* 59, 1–14.

Neumann, W.L. (2014). *Social research methods. Qualitative and quantitative approaches.* (7th ed.). Harlow: Pearson.

Newell, A., Simon, H.A. (1976). Computer Science as Empirical Inquiry: Symbols and Search. *Communications of the ACM, 19*(3), 113–126.

Niehaves, B. (2007). "On Epistemological Diversity in Design Science: New Vistas for a Design-Oriented IS Research?" In *ICIS 2007 Proceedings.* 133.

Nottingam Trent university (2021). NTU Library. Journal metrics indicators—supporting your decision of where to publish. Retrieved from https://www.ntu.ac.uk/__data/assets/pdf_file/0035/865709/journal-metric-indicators.pdf

Offermann, P., Blom, S., Schoenherr, M., Bub, U. (2010). Artefact Types in Information Systems Design Science—*A Literature Review. In Winter,R., Zhao,J.L., Aier, S. (2010). DESRIST 2010, LNCS 6105,* 77–92. Berlin: Springer-Verlag.

Österle, H., Becker,J., Frank, U., Hess, T., Karagiannis, D., Krcmar, H., Loos, P., Mertens, P., Oberweiss, A., Sinz, J. (2011). Memorandum on design-oriented information systems research. *European Journal of Information Systems,* 20, 7–10.

Oswald, G. (2003). *SAP Service und Support.* Germany: Galileo Press GmbH.

Otte, B., Schmidt, E. (2008): Ausbildungsziele, -inhalte und Umsetzungsformen einer prozessorientierten Ausbildung. In Bednarz, S., Schmidt, E., *Arbeitsprozessorientierte und gendergerechte IT-Ausbildung* (pp. 23–37). Bonn, Germany: W. Bertelsmann Verlag Gmbh & Co. KG.

Oxford English Dictionary (2020). *UK Dictionary.* Retrieved from https://www.lexico.com/definition/knowledge.

Paa, L. (2014): *ERP End-User Training mittels e-learning: Kosten, kritische Erfolgsfaktoren und Lernerfolg,* (Unpublished doctoral thesis), University of Innsbruck, Austria.

Paa, L., Ates, N. (2013): Critical success factors of e-learning scenarios for ERP end-user training. *In: Piazolo, F. and Felderer, M. (eds.) Innovation and Future of Enterprise Information System,* 87–100. Berlin,Heidelberg: Springer Verlag.

Paa, L., & Piazolo, F. (2014). ERP-End-User Training through E-Learning: What Should the User Focus on? In F. Piazolo & M. Felderer (Eds.), Novel Methods and Technologies for Enterprise Information Systems, 147–160. Cham: Springer Verlag.

Paape, B., Kiereta, I., Maus, C. (2013). *Wirtschaftsdidaktik. Eine Einführung unter besonderer Berücksichtigung von Handlungs- und Lernfeldorientierung.* Germany: Shaker Verlag.

Paechter, M., Maier, B., Macher, D. (2009). Students' expectations of, and esperiences in E-learning: Their relation to learning achievements and course satisfaction. *Computers and Education*, 54, 222–239.

Pakinee, A., Puritat, K. (2020). Designing a gamified e-learning environment for teaching undergraduate ERP course based on big five personality traits. *Education and Information Technologies*, 26, 4049–4067.

Parush,H., Hamm, H., Shtub, A. (2002). Learning histories in simulation-based teaching: the effects on self-learning and transfer. *Computers & Education*, 39, 319–332.

Patzak, G., Rattay, G. (2014). Projektmanagement. (6 Ed.)., Vienna, Austria: Linde Verlag.

Peterßen, W. (1983): *Lehrbuch Allgemeiner Didaktik.* Germany: Oldenburg Verlag.

Peterßen, W. (2001): *Lehrbuch Allgemeiner Didaktik.* Germany: Oldenburg Verlag.

Petticrew, M., Roberts, H. (2006). *Systematic Reviews in the Social Sciences. A Practical Guide.* USA: Blackwell Publishing.

Piazolo, F., Felderer, M. (2014): *Novel Methods and Technologies for Entreprise Informations Systems.* Berlin, Heidelberg, Germany: Springer Verlag.

Piazolo, F., Felderer, M. (2013): *Innovation and future of Enterprise Information Systems.* Berlin, Heidelberg, Germany: Springer Verlag.

University Duisburg-Essen (2015). *Lern-Psychologie.* Retrieved March 27, 2019, from http://www.lern-psychologie.de/

Plöger, W. (1999): *Allgemeine Didaktik and Fachdidaktik.* Germany: Wilhelm Fink Verlag.

PMA, Project Management Austria. (2018). PM Baseline 3.1. Retrieved July 29, 2022, from https://www.pma.at/de/service/downloads

PMA (2022). Ueber uns. Retrieved July 29, 2022, from https://www.pma.at/de/ueber-uns

Popp, W. (1976): *Kommunikative Didaktik. Soziale Dimension des didaktischen Feldes.* Germany: Beltz Verlag.

Preuss, P. (2017). *In-Memory-Datenbank SAP HANA.* Germany: Springer-Gabler Verlag.

Radianti J., Majchrzak T.A., Fromm, J., Wohlgenannt, I. (2020). A systematic review of immersive virtual reality applications for higher education: Design elements, lessons learned, and research agenda. *Computer & Education*, 147, 1–29.

Raimes, A. (2008). *Keys for writers,* USA: Houghton Mifflin Verlag.

Redaktionsteam PELe, (2006): *Allgemeine Didaktik.* Retrieved from e-teaching.org website https://www.e-teaching.org/didaktik/theorie/didaktik_allg/DidaktischeModelle.pdf

Reinhold, G., Pollak, G., Heim, H. (1999). *Pädagogik-Lexikon.* Austria: Oldenburg.

Reinmann, G. (2005). Innovation ohne Forschung? Ein Plädoyer für den Design-Based Research-Ansatz in der Lehr-Lernforschung. *Unterrichtswissenschaft, 33*(1), 52–59.

Reinmann, G. (2015). *Design-based Research. Erscheint in D. Schemme & H. Novak (Hrsg.), Gestaltungsorientierte Forschung in Innovations- und Entwicklungsprogrammen—Potenzial für Praxisgestaltung und Theoriebildung.* Germany: Bundesinstitut für Berufsbildung.

Reinmann, G. (2018). *Design-Based Research am Beispiel hochschuldidaktischer Forschung.* Retrieved from Gaby Reinmann Hochschuldidaktik website: http://gabireinmann.de/wp-content/uploads/2016/11/Vortrag_Berlin_Nov2016.pdf.

Reinmann, G., Mandl, H. (2006): Unterrichten und Lernumgebung gestalten. *In Krapp, A., Weidenmann, B., Pädagogische-Psychologie. Ein Lerhbuch (5 ed.).* Germany: Beltz PVU, 613–658.

Retzmann, T. (2013). Leitfaden zur Anfertigung eines Unterrichtsentwurfes. Retrieved from the University Duisburg Essen website: https://www.wida.wiwi.unidue.de/fileadmin/fileupload/BWL-WIDA/PDF-Dokumente/Leitfaden_Unterrichtsentwurf_2013-03.pdf

Riedl, A. (2010). *Grundlagen der Didaktik.* Germany: Franz Steiner Verlag.

Ridley, D. (2012). *The Literature Review. A Step-by-Step guide for Students.* (2 Ed.). London: Sage.

Rimmelspacher, U. (2014). *Vertriebsprozesse mit SAP ERP: Mit vollständig integrierten Übungen im Anwendungsmenü und Customizing von SAP ECC 6.0.* Germany: Springer Wieweg.

Roca, J., Chiu, C., Martínez, F. (2006). Understanding e-learning continuance intention: An extension of the Technology Acceptance Model. *International Journal of Human-Computer Studies, 64* (8), 683–696.

Rohs, M., Mattauch, W. (2001). *Konzeptionelle Grundlagen der arbeitsprozessorientierten Weiterbildung in der IT-Branche.* Retrieved from the Publication Database of the Fraunhofer-Gesellschaft website: http://publica.fraunhofer.de/keywords/Arbeitsprozessorientiert.

Rosenshine, B. and Stevens, R. (1986). Teaching functions. *In M. C. Wittrock (Ed.), Handbook of research on teaching,* 3rd ed., 376–391.

Rowe, F. (2014). What literature review is not: Diversity, boundaries and recommendations. *European Journal of Information Systems, 23*(3), 241–255.

Sander, E. (2003): *Didaktische Dimensionen als Instrument der Einordnung und Beurteilung von E-learning-Programmen* (Diplomarbeit). Otto-Friedrich-Universität Bamberg, Germany.

SAP (2014): *TERP10. SAP ERP: Integration of Business Processes.* Germany: SAP AG.

SAP (2022): *TS410. Integrated Business processes in SAP S/4 HANA.* Germany: SAP AG.

SAP Press. (2017). Retrieved December 12, 2017, from https://www.sap-press.com/

SAP Training. (2017). Retrieved December 12, 2017, from https://training.sap.com/

SAP (2019). Retrieved August 05, 2919, from https://training.sap.com/learninghub

SAP (2022). *TS410. Integrated Business processes in SAP S/4 HANA.* Germany: SAP AG.

Sarferaz, S. (2022). *Compendium on Enterprise Resource Planning. Market, Functional and Conceptual View based on SAP S/4 HANA.* Switzerland: Springer Nature Switzerland AG.

Saunders, M., Lewis, P., Thornhill, A. (2009). *Research methods for business students.* (5th Ed.). Harlow: Pearson Education.

Scheibler, J. (2002): *Vertrieb mit SAP Prozesse, Funktionen, Szenarien.* Germany: Galileo Press GmbH.

Scherrer, E., Schaffner, D. (2003): *SAP-Training: Konzeption, Planung und Realisierung (SAP PRESS).* Bonn: Galileo Press GmbH.

Schittler, P. (2001): *Knowledge Management bei ERP-Produkten: dargestellt anhand der Internetplattform der SAP AG* (Unpublished master thesis), University of Vienna, Austria.

Schmitz, M.O. (2022): *Compliance program.* Retrieved June 17, 2022, https://geowiki.lgs-net.com/HGS/Processes/ConformityToLegalRequirements.

Schneckenburger, M. (2005). *Konzeption und Realisierung eines Wissensportals für Hochschulen: eine Implementierung mit mySAP Enterprise Portal.* Germany: Shaker Verlag.

Schotz, B., Kapeso, M. (2014). An m-learning framework for ERP systems in higher education. *Interactive Technology and Smart Education, 11*(4), 287–301.

Scott, J.E., Walczak, S. (2009). *Cognitive engagement with a multimedia ERP training tool: Assessing computer self-efficacy and technology acceptance.* Information & Management, 46, 221–232.

Schryen, G., Benlian, A., Rowe, F., Gregor, S., Larsen, K., Petter, S., Paré, G., Wagner, G., Haag, S., and Yasasin, E. (2017). Literature Reviews in IS Research: What Can Be Learnt from the Past and Other Fields?. *Communications of the Associations Information Systems, 41(30),* 759–774

Schwade, F., Schubert, P. (2016). *The ERP Challenge: An Integrated E-Learning Platform for the Teaching of Practical ERP Skills in Universities.* Procedia Computer Science 100, 147–155.

Schwerer, F., Egloffstein, M. (2016). *Participation and achievement in enterprise MOOCs for professional learning.* 13th International Conference on Cognition and Exploratory Learning in Digital Age.

Seethamraju, R., (2011). Enhancing student learning of enterprise integration through ERP business simulation game. *Journal of Information Systems Education, 22*(1), 19–29.

Seta, B.H., Wati, T., Muliawati, A., Hidayanto, A.N. (2018). E-learning Success Model: An Extention of DeLone & McLean IS' Success Model. *Indonesian Journal of Electrical Engineering and Informatics (IJEEI), 6* (3), pp. 281–291.

Seufert, S., Back, A., Häusler, M. (2001). *E-learning—Weiterbildung im Internet: Das «Plato-Cookbook» für internetbasiertes Lernen.* Kilchberg, Germany: SmartBooks.

Shah, V.S. (2015). *Essays on Technology-Mediated Training: Implications for Design and Evaluation* (Dissertation, Texas A&M University, USA). Retrieved from http://scholarwo rks.uark.edu/etd/1347.

Simon, A. (1996). *The science of the artificial.* England: The MIT press.

Sonnenberg, C., & vom Brocke, J. (2012). Evaluation Patterns for Design Science Research Artefacts. In M. Helfert & B. Donnellan (Eds.), *Proceedings of the European Design ScienceSymposium (EDSS)* 286, pp. 71–83. Dublin, Ireland: Springer Berlin/Heidelberg.

Stacie, P., DeLone, W., McLean, E. (2008). Measuring Information systems success: models, dimensions, measures, and interrelationships. *European Journal of Information Systems, 17* (3), pp. 236–263.

Stenberg, R.J., Stenberg, K. (2012). *Cognitive Psychology.* Belmont, USA: Wadsworth

Stracka, A.G., Macke, G. (2002). *Lern-Lehr-theoretische Didaktik.* Germany: Waxmann Verlag GmbH.

Sun, P.C., Tsai, R.J., Finger, G., Chen, Y.Y., Yeh, D. (2008). What drives a successful e-learning? An empirical investigation of the critical factors influencing learner satisfaction, *Computers & Education,* 50, pp. 1183–1202.

Terhart, E. (2009). *Didaktik. Eine Einführung.* Stuttgart: Philipp Reclam jun. .

Teufel, T., Röhrricht, J., Willems, P. (2000). *SAP-Prozesse mit knowledge maps analysieren und verstehen [business engineering, Knowledge-Management, mind mapping und enjoySAP für die Grundlagenprozesse von mySAP.com]*. Germany: Addison-Wesley Verlag.

Thomas, J., Harden, A, Newmann, M. (2012). Synthesis: combining results systematically and appropriaptely. In: D. Gough, S., Oliver, and Thomas, J. *An introduction to systematic reviews*, 179–226. London: Sage.

Tuckman, B.W., Harper, B.E. (2012). *Conducting Educational Research*. UK: Rowman & Littlefield Publishers, Inc.

University of Augsburg. Institut für Medien und Bildungstechnologie. (2017). *Design Based Research*. Retrieved from http://qsf.e-learning.imbuniaugsburg.de/node/540.

University Duisburg-Essen (2015). *Lern-Psychologie*. Retrieved March 27, 2019, from http://www.lern-psychologie.de/

University of Education Ludwigsburg (n.d.). *Planung von Unterricht*. Retrieved from (https://www.ph-ludwigsburg.de/fileadmin/subsites/2b-dtsc-t-01/user_files/gans/material/unterrichtsplanung.pdf

Uljens, M. (2005): *School didactics and learning. A school didactic model framing an analysis of pedagogical implications of learning theory*. UK: Psychology Press, Ltd.

Vaishnavi, V., Kuechler., B. (2015). *Design Science Research Methods and Patterns: Innovating Information and Communication Technology*. (2nd Ed.). New York: CRC Press.

Vaishnavi, V., Kuechler., B. & Petter, S. (2018). *Design Science Research in Information Systems*. Retrieved from Design Science Research in Information Systems and Technology website: http://desrist.org/knowledge.

Van Aken, J.E. (2004). Management research based on the paradigm of the design sciences: The quest for field-tested and grounded technological rules. *Journal of Management Studies*, 41, 219–246.

Venkatesch, V., Davis D., F. (2000). *A Theoretical Extension of the Technology Acceptance Model: Four Longitudinal Field Studies*. Management Science *46* (2), 84–204.

Verband der Hochschullehrer für Betriebswirtschaft e.V. (2021). *VHB-Jourqual, VHB-Jourqual 3*. Retrieved November 21, 2021, from https://vhbonline.org/vhb4you/vhb-jourqual/vhb-jourqual-3.

Wahl, M. (2003): *Wissensmanagement im Lebenszyklus von ERP-Systemen : explorative Untersuchung und Entwicklung eines Gestaltungskonzeptes für SAP R/3-Projekte*. Germany: Fachverlagsgruppe Beterlsmann Springer.

Warwitz, S., Rudolf, A.: *Projektunterricht. Didaktische Grundlagen und Modelle*. Germany: Verlag Hofmann.

Watson, J.B. (1997). *Behaviourismus*. Germany: Dietmar Klotz Verlag.

Webster, J., Watson, R.T. (2002). Analysing the past to prepare for the future. *MIS Quarterly*, 26(2), xiii-xxiii.

Weiten, W. (2008). *Psychology: Themes and Variations*. USA: Thomson Learning, Inc.

Wikipedia. (2022). Chamilo. Retrieved July 7, 2022, from https://de.wikipedia.org/wiki/Chamilo.

Wikipedia (2022). PHP. Retrieved July 7, 2022, from https://de.wikipedia.org/wiki/PHP

Winkel, R. (1999). Die kritisch-kommunikative Didaktik. In Gudjons, H., Winkel, R. (Ed.), *Didaktische Theorien (*pp. 93–112*)*. Hamburg, Germany: Bergmann und Helbig.

Wopp, C. (1986). *Unterricht, handlungsorientierter*, in: Haller, H.-D., Meyer, H., Ziele und Inhalte der Erziehung und des Unterrichts, Enzyklopädie Erziehungswissenschaften, Germany: Beltz Juventa.

Wu, J.H., Tennyson, R.D., Hsia, T.L. (2010). A study of student satisfaction in a blended e-learning system environment. *Computers & Education*, 55, pp. 155–164.

Yahia, S., Arshad, N. H. (2009). E-Education systems implementation success model. *International Journal of Education and Development using Information and Communication Technology (IJEDICT), 2009, 5* (2), pp. 123–133.

Zhao, Y.A., Srite, M., Kim, S., Lee, J. (2021). Effect of team cohesion on flow: An empirical study of team-based gamification for enterprise resource planning systems in online classes. Decision Sciences, Journal of innovative education, 19, 173–184.

The manufacturer's authorised representative in the EU is Springer
Nature Customer Service Centre GmbH, Europaplatz 3, 69115 Heidelberg,
Germany. If you have any concerns regarding our products, please
contact ProductSafety@springernature.com

Printed and bound by CPI Group (UK) Ltd, Croydon, CR0 4YY
28/04/2026
02098512-0001